Evolutionary Concepts in Immunology

Robert Jack • Louis Du Pasquier

Evolutionary Concepts
in Immunology

 Springer

Robert Jack
Department of Immunology
University of Greifswald
Greifswald, Germany

Louis Du Pasquier
Department of Environmental Sciences, Zoology
University of Basel
Basel Stadt, Switzerland

ISBN 978-3-030-18669-2   ISBN 978-3-030-18667-8   (eBook)
https://doi.org/10.1007/978-3-030-18667-8

This Springer imprint is published by the registered company Springer Nature Switzerland AG.
The registered company address is: Gewerbestrasse 11, 6330 Cham, Switzerland

Preface

Dobzhansky's slogan, "Nothing in biology makes sense except in the light of evolution" [1], applies with special force to immunology—a complex subject that many students find hard to come to grips with. It is for them that this book was written. It is neither a textbook of immunology, nor is it a textbook of evolution. Rather, it is an attempt to show how evolutionary forces have shaped immune systems across phylogeny.

In this enterprise, we are caught on the horns of the old dilemma: explain "more and more about less and less" until one ends up presenting everything about nothing—or go instead for "less and less about more and more" until one ends having said nothing about everything. We have chosen the dangerous middle road. We make no attempt to be comprehensive. Instead, we have chosen to present certain immune phenomena in an evolutionary context—well aware that the choice of phenomena handled reflects little more than our personal preferences. We do, however, attempt to broaden the perspective by summarising a larger range of examples in the appendices. These examples are intended to indicate the breadth of the subject and to provide key words for those who wish to dig a little deeper. In a similar vein, we do not cite all the important papers that provide the data that we discuss. Our citations are restricted to a few "classic" papers and to well-written, accessible reviews. As with the appendices, they are there to provide a starting point for those who wish to find out more about particular issues.

Greifswald, Germany   Robert Jack
Basel Stadt, Switzerland   Louis Du Pasquier

Reference

1. Dobzhansky T (1973) Nothing in biology makes sense except in the light of evolution. Am Biol Teach 35:125–129

Acknowledgements

We thank Claudia Berek, Katja Simon, Clara Simon, Philipp Enghard and Daniela Brites for critical comments on the manuscript.

Contents

1   What Makes Evolution Tick?   1
   1.1   "How" and "Why" in Biology   1
   1.2   Species: The Importance of Variation   2
   1.3   Species: The Importance of Selection   3
   1.4   The Concept of Fitness   3
   1.5   Genes   4
   1.6   Mutations and Variants   6
      1.6.1   Mutations in Soma and Germline   6
      1.6.2   Functional Classes of Mutations   7
   1.7   Genetic Drift   9
   1.8   Recombination   11
   1.9   Host–Pathogen Interactions Are an Arms Race   13
   1.10 The "Generation Gap"   13
   1.11 Evolution of Immunity   14
   References   16

2   From Unicellular to Metazoan Immunity   17
   2.1   Germline and Soma   18
   2.2   Phagocytosis: An Old Habit Becomes Restricted   19
      2.2.1   Lineage Restriction of Phagocytic Competence
         in Metazoans   19
      2.2.2   Professional Phagocytes Exploited by Pathogens   20
      2.2.3   Pathogens Can Reactivate and Exploit Phagocytic
         Competence   21
      2.2.4   Diversification of Phagocytic Competences
         in Metazoans   21
      2.2.5   Autophagy: A Eukaryotic Cousin of Phagocytosis   22
   2.3   Metazoans Are Societies of Cells: The Importance of "Self"   23
      2.3.1   Recognition of "Apoptotic-Self"   23
      2.3.2   Recognition of "Necrotic-Self"   24

      2.3.3   When Cooperation Is Not Enough: Recognition
              of "Oncogenic-Self" 25
  2.4   Immobile "Epithelial" Immunity in Basal Invertebrates 26
  2.5   Immune Defence in Complex Metazoans Must Be Mobile 27
  2.6   Two Arms of Immunity: "Innate" and "Adaptive" 28
      2.6.1   The "Adaptive Problem" in Vertebrates: How Many
              Specificities? 29
      2.6.2   Metchnikoff's Legacy 29
      2.6.3   Phagocytosis and Janeway's "Dirty Little Secret" 30
  References 31

3   Innate Immunity 33
  3.1   Structure of the Innate Immune System 33
  3.2   Evolution of Innate Immune Receptors 34
      3.2.1   Gene Duplication 36
  3.3   Evolution of a Gene Family by Duplication and Modification:
      the Toll-Like Receptors 36
      3.3.1   *Drosophila* Toll-1 38
      3.3.2   Mammalian Toll-Like Receptors 39
  3.4   Exons and Introns: Making New Proteins by Swapping
      Domains 40
      3.4.1   Exon Shuffling: Speeding Up the Rate of Acquisition
              of New Genes 40
      3.4.2   Exon Splicing: One Gene—More Products 42
      3.4.3   Adaptive Introgression 42
  3.5   Extracellular Innate Immune Receptors and Their Targets 43
      3.5.1   Receptors that Target Surface Structures
              on Extracellular Microbes 44
      3.5.2   Receptors that Target Prokaryotic Surface Molecular
              Patterns 46
      3.5.3   Provision of Endogenous Bait 47
  3.6   Intracellular Innate Immune Receptors and Their Targets 47
      3.6.1   Sensing the Location of Nucleic Acids: APOBEC3G ... 48
      3.6.2   Sensing the Local Concentration of DNA: cGAS
              and STING 49
      3.6.3   Sensing RNA Fine Structure: RIG-I 50
      3.6.4   Detecting Intracellular Bacterial Pathogens 51
      3.6.5   Detecting Endogenous "Danger" Signals 52
      3.6.6   "Missing-Self" and the Evolution of "Natural Killer Cell"
              Receptors 53
  3.7   Living with Commensals 54
  3.8   Beyond Receptor-Ligand Interactions: Immune Systems
      as Computational Devices 55
      3.8.1   Signal Transduction 55
  3.9   The Outputs 58

3.9.1 Cell Movement and the Inflammatory Response 58
3.9.2 Complement-Mediated Phagocytosis 61
3.9.3 Programmed Cell Death 66
3.10 Who Needs More? 67
References 68

4 The Triumph of Individualism: Evolution of Somatically Generated
Adaptive Immune Systems 71
4.1 Adaptive Systems that Use Nucleic Acid Sensors 72
4.1.1 Bacterial CRISPR-Cas 72
4.1.2 Eukaryotic RNA Interference 73
4.1.3 Evolution of Adaptive Immune Systems Based
on RNA Interference 73
4.2 Somatic Evolution of Immune Systems that Use Protein
Sensors 75
4.2.1 Two Different Protein-Based Adaptive Immune
Systems 76
4.2.2 Lymphocytes and Their Receptors 77
4.3 Somatic Formation of the Agnathan Adaptive Immune
Receptors 78
4.3.1 The Key Enzyme: An AID-Like Cytidine Deaminase . . . 79
4.3.2 Structure of the Receptor Molecules 79
4.3.3 Tolerance 80
4.4 Adaptive Immune Receptors in Gnathostomes 81
4.4.1 The Immunoglobulin Super Family Domain 81
4.4.2 The "Transib" Transposon Contributed to the Structure
of All Antigen-Specific Receptors in Jawed
Vertebrates 81
4.4.3 The Gnathostome Immune Receptor Antigen-Binding
Sites 83
4.5 RAG and Its Limitations 84
4.5.1 Somewhere Between Lots of Receptors and Lots
of Junk 86
4.5.2 RAG Recombination: A Half-Hearted Attempt to Build
an All-Encompassing Adaptive Repertoire 86
4.5.3 Ligand Binding and "Tolerance" in the B-Cell
Lineage 87
4.5.4 Allelic Exclusion 87
4.6 Ligand-Binding in the T-Cell Lineages 88
4.6.1 The MHC Complex and the Peptide Carrier
Molecules 89
4.6.2 Why Two Types of MHC-Peptide Carrier
Molecules? 91
4.6.3 TCRs Recognise Complexes of Peptides and MHC
Molecules 93

4.7   Which TCRs Are Potentially Useful: "Positive Selection"   93
4.8   "Negative Selection" of T-Cells in the Thymus   94
        4.8.1   Presenting All "Self" Peptides   95
        4.8.2   Expressing "Self" Peptides on MHC Class-II   95
4.9   The Price of T-Cell Selection in the Thymus   96
4.10   Co-evolution of TCRs and MHC-Molecules   96
4.11   Repertoire Change After Central Tolerance: Somatic
        Hypermutation of BCRs   97
4.12   Life After Central Tolerance: Peripheral Tolerance
        and Lymphocyte Activation   99
        4.12.1   The "Two Key" Fail-Safe Strategy   100
        4.12.2   T-Regulator Cells   102
4.13   Beyond the Receptor Repertoire: Lymphocyte Effector
        Functions   102
4.14   Adaptive Memory in Gnathostomes   103
4.15   Diversity of Adaptive Immune Repertoire Formation
        in Gnathostomes   104
4.16   The Evolutionary Relationship of the Agnathan
        and Gnathostome Adaptive Immune Systems   106
        4.16.1   Homology and Analogy   107
        4.16.2   Intercalary Evolution and "Deep Homology"   109
        4.16.3   The Adaptive Niche   110
        4.16.4   Haematopoiesis and the Origin of Lymphocytes   110
        4.16.5   Thymus and "Thymoid"   112
        4.16.6   Evolution of AID-Like Cytidine Deaminase Functions
                  in Immunity   112
        4.16.7   The Last Common Ancestor of Agnathans and
                  Gnathostomes   114
References   115

5   The Other Side of the Arms Race   119
5.1   Coming to Terms with Pathogens   120
5.2   Pathogen Strategies Directed Against the Innate Immune
        System   122
        5.2.1   Hiding from the Innate Receptor Repertoire   122
        5.2.2   Interfering with the Innate Response   124
5.3   Examples of Pathogen Strategies to Avoid Adaptive
        Immunity   125
        5.3.1   Disrupting Adaptive Immunity's Detection Hardware . . .   125
        5.3.2   Playing Dead   126
        5.3.3   Hiding in the Immunological Future   128
References   130

6   Postface   131

Appendices 133
  Appendix A: A Simplified Classification of Metazoa 133
  Appendix B: Immune Receptors and Their Common Domains 135
  Appendix C: cGAS, STING and Cyclic Dinucleotides 136
  Appendix D: Immunoglobulin Superfamily Domain 137
  Appendix E: Intercalary Evolution 138
  References 142
Index 143

List of Abbreviations

AID	Activation-induced deaminase
APC	Antigen presenting cell
APOBEC3G	Apolipoprotein B mRNA editing enzyme catalytic subunit 3G
BCR	B-cell receptor
C1q	Complement factor 1q
C3	Complement factor 3
CAMP	Cationic antimicrobial peptide
CDR	Complementarity determining region
cGAMP	cyclic guanine-adenine mono phosphate
cGAS	cyclic guanine-adenine dinucleotide synthase
CRP	C-reactive protein
dC	deoxy cytidine
dU	deoxy uridine
FREP	Fibrinogen-related proteins
GPI	Glycosylphosphatidylinositol
HCMV	Human cytomegalovirus
HIV	Human immunodeficiency virus
IgNAR	Immunoglobulin new antigen receptor
IgSF	Immunoglobulin superfamily
LMP	Large multifunctional protease
LPS	Lipopolysaccharide
LRR	Leucine-rich repeat
MAC	Membrane attack complex
MAMP	Microbe-associated molecular pattern
MASP	MBL-associated serine protease
MBL	Mannose-binding lectin
MHC	Major histocompatibility complex
miRNA	micro RNA
NK cell	Natural killer cell
NLR	NOD-like receptor
NOD	Nucleotide-binding oligomerisation domain-containing protein

PAMP   Pathogen-associated molecular pattern
PAX5   A transcription factor
PAX6   A transcription factor
R_o Epidemiological basic reproduction number
RAG   Rearrangement-associated gene
RIG-I   Retinoic acid-inducible gene 1
RISC   RNA-induced silencing complex
STING   Stimulator of interferon genes
TCR   T-cell receptor
TEP   Thio-ester containing protein
TLR   Toll-like receptor
VLR   Variable lymphocyte receptor
VSG   Variable surface glycoprotein
WGD   Whole genome duplication

Chapter 1
What Makes Evolution Tick?

1.1 "How" and "Why" in Biology

In this book we will be looking at some of the evolutionary forces that shaped, and are shaping, immune defence systems. However, before doing that it makes sense to look at some of the basic features of the evolutionary process. Perhaps the best place to start is with an essay written by Ernst Mayr, at that time professor of Zoology at Harvard, which was published in Science in 1961 [1]. In this essay Mayr considered the case of a little bird that fluttered around outside his office window all summer and then, as autumn approached, took off and flew south. Mayr pointed out that there are two types of question one might ask about all this. The first is what he called the "how" question. How does the little bird know that the year is nearly over? How does it organise its flight south? Clearly the bird must be able to detect some feature of approaching autumn—perhaps by using some appropriate receptor system to detect the shortening of the day or the gradual fall in average daily temperature. Then it must convert this signal into electrical impulses that can be interpreted by its brain, after which an output signal must be generated, which encourages the little bird to fly off in the correct direction. Finding the answers to the "how" questions would involve a lifetime or two of fascinating research, and the results would be expressed in terms of molecules and of biochemical and neural pathways. These things are familiar to all of us, since molecules and pathways are what most scientists spend most of their working lives dealing with.

Mayr pointed out, however, that there is a second type of question, which might be asked, and that is the "why" question. Why does the little bird fly south? The answer, in general terms, is that it flies south because it is an insect-eating bird and there aren't a lot of insects flying around in Massachusetts in the winter. We are dealing therefore with a case of "survival of those who fly south". It is a matter of Darwinian evolution the basic components of which are variation due to random mutation, genetic drift, recombination and natural selection. "How" questions are concerned with the ways in which the information stored in the genome is expressed

© Springer Nature Switzerland AG 2019

R. Jack, L. Du Pasquier, *Evolutionary Concepts in Immunology*,

https://doi.org/10.1007/978-3-030-18667-8_1

and made use of. "Why" questions, in contrast, have to do with the way the genome became the way it is. Any good recent textbook of immunology will supply answers to many of the "how" questions. Our aim is instead to provide an introduction to the "why" questions behind immune defence.

1.2   Species: The Importance of Variation

During the nineteenth century—and beyond—one of the principle jobs of biologists was to systematise living things and assign them to distinct species. At that time, the principle tool available was morphology. The problem was, however, that one man's decisive morphological difference might be regarded by another as a mere detail. Not surprisingly, bitterly fought arguments arose on the question of species assignments and on the status of species, varieties and races. Many an academic career was made or broken on these questions.

Charles Darwin's experiences on the Beagle voyage led him to cut through this Gordian knot by realising that what was important was not the "sameness" of the members of a species, but rather the fact that each member was a variant of a common theme. He was not the first to come to this conclusion, but what made it into an epochal breakthrough was his realisation that some of these variants must be better suited to their environment than others, and thus would fare better in the struggle for existence. Nevertheless, this realisation that all members of a species are variants on a common theme came with a price. The man who wrote the seminal work on the origin of species had to tell us at the start of his book that he didn't really have any precise demarcation criteria to tell us what is—and what is not—a species. In Chap. 2 of the "Origin of Species" he wrote:

> ... I look at the term species, as one arbitrarily given for the sake of convenience to a set of individuals closely resembling each other, and that it does not essentially differ from the term variety, which is given to less distinct and more fluctuating forms. The term variety, again, in comparison with mere individual differences, is also applied arbitrarily, and for mere convenience sake.

Attempts to supplement the morphological species definition with one based on the ability of two individuals to produce fertile offspring fared little better, for though interspecies crosses are frequently sterile there are many exceptions. Darwin summed this up as follows:

> Finally, looking to all the ascertained facts on the inter-crossing of plants and animals, it may be concluded that some degree of sterility, both in first crosses and in hybrids, is an extremely general result; but that it cannot, under our present state of knowledge, be considered as absolutely universal.

Sterility in crosses is thus not a generally applicable criterion, which can be used to define the term species.

Despite the vast progress in biology since then, there is not much more that can be said about the definition of the word "species". It was—and remains—one of that curious set of words, which everybody more or less understands, but which nobody can precisely define. The lack of a precise definition is, however, not a major impediment, and it pales into insignificance against Darwin's recognition that variation in character between individuals is one of the central elements in evolution.

1.3   Species: The Importance of Selection

Variation between the members of a population is one of the keys to evolution, but variation on its own is of no value. Variation is only useful in situations in which natural selection is able to choose which of the variants will best contribute to future generations. One sees this, for example, in the case of the Coelacanth, a deep-water fish, which is often referred to as a "living fossil", since it looks just like ancient members of the species present in the fossil record from 400 million years ago. Of course this is merely a journalistic simile since the genome of the fish, like that of every other organism, has been accumulating mutations for all these millions of years. Yet, down in the depths of the sea, not much in the way of environmental change has taken place, and so there has been little if any selective pressure to force adaptation to new conditions, and hence to drive the evolution of morphological changes by positively selecting appropriate mutations. Evolution thus depends not only on variation by mutation, but also on selection. Mutations are ten a penny—it is selection that makes mutations evolutionarily significant.

1.4   The Concept of Fitness

The story of Mayr's insect-eating little bird does not end with its flight south, for once there the little bird grew fat on southern insects, found a mate and, in the following spring, flew back north again to found a family. The pair raised a brood of five chicks. All the other little birds were doing the same, so that by late spring the population had exploded. Ever since Malthus's essay on the Principle of Population, published in 1798, we know that this story can't have a happy ending. All these extra birds have to be fed from a limited population of insects, and so some of the chicks will starve, others will be carried off by disease, while predators will fatten themselves on most of the rest. Few of the babies will survive the year and be able to start their own families the next spring.

So who survives?
And who dies?
And why?

The catchy phrase "survival of the fittest", first suggested by the English philosopher Spencer in 1864, has been criticised, because if the fitter survives, then the phrase merely states "survival of the survivors"—which is not informative. The criticism is, however, mistaken, because it fails to appreciate that the term fitness in these discussions always refers to genetic rather than to physical fitness. It is the sequence of the bases in a genome that determines its genetic fitness, and hence the probability that it will be projected into subsequent generations. A genome that is, for whatever reason, better suited to the current environment will have a better chance of being found in the group of survivors. It will almost inevitably be a winner in what Darwin referred to as the "struggle for existence".

In many cases the solutions to problems facing organisms involve compromises and trade-offs that have to take account of many different imperatives. For example, the nematode worm *Caenorhabditis elegans* is a self-fertilising hermaphrodite that starts its reproductive life by making sperm, which it stores. Having produced the sperm it then switches to producing eggs. These eggs are fertilised using the stored sperm and, crucially, the number of stored sperm limits the number of offspring the worm can produce. A mutant that makes extra sperm can therefore produce more progeny, and more progeny should translate into higher fitness—but, oddly enough, that is not the case. Nematode worms survive by eating bacteria and their voracious eating habits result in the food supply quickly being exhausted. A successful nematode is one that eats early, and immediately starts producing young. This is where the mutant worm loses, for the time it devotes to producing extra sperm is time that is lost in the race to produce fertilised eggs. The mutant's eggs are laid later than those of its wild type peers, and by the time these mutant eggs hatch their wild type rivals have scavenged the available food. In this case the capacity to produce more progeny does not lead to greater fitness—quite the reverse [2].

Even in more straightforward cases "survival of the fittest" works only for a part of the spectrum of selective pressures and mutation rates. When selective pressure is low, then even modestly deleterious mutations may survive. Equally, when the selection pressure suddenly becomes excessive, as happened during the massive environmental disruptions at the end of the Permian, which led to the extinction of over 90% of marine species, no survivable rate of mutation can generate appropriate adaptive changes. However, between these extremes, the ability of a species to survive changes to its environment for any significant length of time is determined by natural selection sifting through the variants present in the population. The genetically fitter alleles, which have accumulated mutations beneficial in the current environment, will be successful. Those that fail to accumulate the necessary beneficial mutations will disappear.

1.5   Genes

One of the greatest problems for evolutionists in the nineteenth century was that nobody understood anything about the mechanisms of heredity. The nature of the genetic information was a mystery, and the power of random mutation to alter it was also completely unknown. Nowadays that has all changed, because we know a great deal about DNA and chromosomes and heredity. We have clear ideas on how the information in the DNA is converted into the proteins that carry out the biochemical work of the cell, and of how changes in the DNA can alter the structure and expression of these proteins. We also know quite a bit about how mutations, by inducing changes in the information encoded in the DNA, can alter the functionality of this central molecule of heredity.

It's not hard to imagine the potentially disastrous consequences of changing the genetic information that defines the structure of proteins. However, mutations are induced randomly (more or less) across the genome, and since only around 2% of all the bases in the DNA of a human chromosome directly code for proteins, it follows that most mutations affect non-coding regions of the genome. That, however, does not mean that such mutations are necessarily harmless. Though it is by no means yet entirely clear what all of the DNA that does not code for proteins is good for [3], it is clear that many types of RNA molecule with important functions in the cell are encoded here, as are sequences like promoters, enhancers, silencers and locus control regions all of which regulate the extent to which the transcribed sequences are expressed. Mutations that interfere with such sequences may well have drastic deleterious effects.

We have struggled through the last few sentences in a rather inelegant attempt to avoid using the word "gene". The reason for this is very simple. To take account of all the known eventualities, current definitions of the word "gene" have to be couched in terms so broad and vague that they fail to exclude almost any sequences in the genome. "Gene", like "species", is one of those words, which everybody understands, and yet nobody can accurately define. For the purposes of this book we will use the term "gene" in a rather loose way to refer to those sequences in the genome, which are required for the regulated expression of a protein or of RNA.

Analysis of the complete genome sequences from a large number of different animals tells us, to a first approximation, how many genes these organisms contain. The genomes of human beings and mice contain roughly 20,000 genes, as do the genomes of many other animals including the flatworm *Macrostomum lignano*. Is 20,000 a large number? That is a little bit like asking, "How long is a piece of string?" It all depends on your point of view. But you can get some sort of answer by looking in the mirror, and asking whether you would be able to build the human being you see there using just 20,000 genetic instructions. These 20,000 genes have to build our cells with all the complex metabolic processes going on inside them. They have to organise these cells into tissues and organs. They have to make you able to sense your environment and provide you with the ability to respond optimally to unexpected events. With this sort of job description, 20,000 genetic instructions is truly not a lot. In reality, life is run on a genetic shoestring.

Since flat worms and humans have roughly the same number of genes, many of which perform the same functions in both organisms, it is clear that an increase in the complexity of an animal is not achieved by an increase in the number of genes, but rather by an increase in the extent of integration and co-ordination of the expression and functions of the genes we have. Integration, rather than numbers, is the name of the game.

1.6   Mutations and Variants

More than one and a half centuries after Darwin, we now know that the variations in phenotype, which led him to his breakthrough, are due to spontaneous random mutations that result in the alteration of the structure of DNA. The term mutation has a bad press, for it conjures up thoughts of cancer, destruction and death, but mutation is, in addition to all of that, one of the central driving forces of evolution. Mutations are happening all the time, so that all genomes are constantly acquiring new mutations—a process that is known as "genetic drift". So long as the mutations are not frankly deleterious they may be tolerated, at least for a while, and this genetic drift is the source of the variation that provides the choices for natural selection to work with.

Where do the mutations come from? There are numerous sources among which are, for example, the errors that inevitably happen during the processes of replication, when the chromosomes are duplicated prior to cell division. This replication process is astonishingly precise—but nothing in life is perfect and, with 6,000,000,000 bases of new DNA to be synthesised each time a diploid human cell divides, the odd error or two is bound to occur. When one considers that each of us started off life as a single fertilised egg cell and that an adult human being consists of approximately 10^{14} somatic cells, it will be clear that there has been an awful lot of replication in your life and in mine. Even this large number is a serious underestimate, since numerous lineages in the body, including many lymphocytes, myeloid cells and epithelial cells turnover rapidly throughout the course of our life.

In a rather similar way, the hurly burly of transcription, when the information in the DNA is copied into RNA, is potentially mutagenic. So too are extrinsic factors such as exposure to heat, aridity, ionising radiation or to environmental chemicals able to modify nucleic acids. Of course life would not be possible for long without sophisticated systems to detect and correct mutational damage, and all living things possess such error-repair mechanisms. Where necessary, as in the bacterium *Deinococcus radiodurans* or in some arthropods like scorpions and cockroaches, which are able to flourish despite exposure to high levels of radiation, these repair systems can be tuned up and optimised to a remarkable degree.

1.6.1   Mutations in Soma and Germline

It makes a huge difference in which cells the mutations happen. Mutations induced in the "germline", i.e. in those cells which are responsible for the generation of eggs and sperm, are passed on to future generations, and so this sort of mutation becomes the source of inherited variation between individuals. In contrast, mutations induced in normal somatic cells of the body—liver cells, neurons, muscle cells, etc.—may well affect the life of an individual, but they cannot be passed on to the next generation, and hence are generally thought of as being of no evolutionary interest.

Such mutations never become part of the germline encoded genetic future and that is why Lamarck's hypothesis of the heritability of acquired characteristics is—with few exceptions (see Sect. 4.1.3)—a non-starter. That is not to say that somatic mutations have no evolutionary significance, for those causing childhood cancers will dramatically reduce the probability that an affected individual will reproduce, and hence contribute to the genetic future. On the other hand, there is one large class of somatic mutations, which, though not heritable per se, do have a large positive effect on individual fitness, and hence are of considerable evolutionary significance. These are the somatic mutations in the adaptive immune systems of vertebrates that lead to the modification of the antigen-receptors of the lymphocytes (discussed in Chap. 4). This most unusual example of the exploitation of somatic variation provided the vertebrate immune systems with a powerful weapon in the struggle against pathogens. So powerful is this weapon that it is perhaps not surprising that some pathogens have learned to counter it, by themselves developing virulence strategies dependent on somatic mutation [4].

1.6.2   Functional Classes of Mutations

Mutations are frequently classed into three types: deleterious mutations, which will be negatively selected; neutral or silent mutations, which have no effect and hence will be ignored by natural selection; and finally—by far the smallest group—advantageous mutations, which will be positively selected. The simplest way to envisage these mutations is to think of a stretch of DNA, which codes for a protein. This protein must be folded into a precisely defined three-dimensional shape in order to do its job as an enzyme or as a structural element. A mutation, which alters the amino acid sequence of the protein, may well prevent the achievement of the correct shape, and in such cases the protein may lose its capacity to do its job. If that job is crucial for the life of the organism, then the mutation will clearly be deleterious, and natural selection will ensure that this sort of mutation fails to spread widely in the population. However, by no means all mutations, which change the order of the bases in a protein coding sequence, will necessarily lead to a disaster. There are two reasons for this. The first is that for most proteins some positions in the sequence can accommodate a change of amino acid without any adverse effect on function. Such neutral mutations are essentially invisible to natural selection and hence will be tolerated. The second type of mutation, which is likewise essentially invisible to natural selection, are the so-called silent mutations. These arise because the nucleic acid code is a triplet code in which 3 bases in the messenger RNA define each amino acid in the protein. This leads to a problem, because there are 64 possible triplets, or "codons", but only around 20 amino acids. The code is therefore degenerate, in the sense that many of the amino acids have more than one codon. For example, the amino acid leucine is coded for by six different triplets, and one third of all of the base changes possible in these triplets will result in the exchange of one leucine codon for another. Such mutations in the protein coding regions, which do not result in the change of an amino acid at that position, will clearly not alter the structure of the protein and hence are usually referred to as silent mutations.

The assertion that such mutations are indeed "silent" is probably true in most cases, but there are two caveats. The first is that because the transfer-RNAs, which are needed to bring the amino acid into position for the protein synthesis apparatus, are not all present in the same concentration in the cell, a silent mutation may result in the use of a rare tRNA in place of an abundant one, and hence change the rate at which the protein chain is synthesised. This, in turn, can lead to adverse effects on the folding of the protein into its proper shape. Recent large scale sequencing efforts looking for evidence of negative selection against the use of rare tRNAs in the human population suggest that in most cases this is not a major problem [5], but there is also evidence that in rare cases silent mutations, in certain crucial positions, may have a significant effect on the ability to produce functional protein in adequate amounts.

The second caveat to the "silence" of such mutations is that they may alter a protein's property of evolvability. Suppose that some protein has, at a particular position, a leucine coded for in the mRNA by the triplet UUA, and that the gene undergoes a single base silent change, so that the codon now reads UUG, which also codes for leucine. Let us further suppose that at some point in the future a change in the environment leads to a situation in which the organism would be much better off with a tryptophan at this position rather than a leucine. Tryptophan is coded for only by the triplet UGG, so the variant, which codes for leucine using the UUG codon, can be readily mutated by a single base substitution to this UGG codon. The variant which codes for leucine using the UUA codon has a problem, because it requires that two random mutations instead of one take place—and that is going to be a much slower process. The variant with UUG can thus be more readily adapted to deal with the changed conditions. The so-called silent mutations are thus by no means always inconsequential alterations to the genome.

The advantageous mutations come last, because this is by far the smallest group, the majority of mutations being either neutral or deleterious. Nevertheless, if an advantageous mutation does arise, then natural selection will favour its spread through the population. How fast this happens will depend on the degree of advantage the mutation brings with it. For example, mutations that overcome the otherwise normal intolerance of adult humans to lactose in milk have arisen independently numerous times in cattle herding societies. Similarly, mutations which confer a degree of resistance to malaria caused by infection with *Plasmodium falciparum* have arisen and been positively selected several times over the last 10,000 years or so. The best studied such mutation is caused by a single base change in the gene coding for the ß chain of haemoglobin, and it results in the substitution of a valine residue in place of the normal glutamic acid residue at position 7 of the polypeptide chain. Those individuals homozygous for the mutation suffer from sickle cell anaemia, a potentially life-threatening condition resulting from the tendency of the mutant haemoglobin chain to aggregate, especially at low oxygen tension. The aggregated haemoglobin deforms the erythrocyte making it liable to clumping and lysis in the circulation and, in addition, these deformed erythrocytes are removed by the erythrocyte quality control system in the spleen. The results are anaemia, as well as a massive overload of the spleen, which can cause the destruction of this organ. In heterozygous individuals, on the other hand, these serious clinical effects are absent because the presence of the normal haemoglobin ß chain ameliorates the tendency of the mutant form to aggregate. Only at very low oxygen pressure, when for example an erythrocyte is infected with the metabolically highly active *Plasmodium falciparum* parasite, which is using up oxygen as if there were no tomorrow, does the erythrocyte suffer from the damaging effects of haemoglobin aggregation. Thus, in the blood stage of malaria infection, the deformed infected erythrocytes tend to be removed early leading to a markedly reduced patient mortality. This example demonstrates the complexities involved in evolutionary interactions, for the substantial advantage gained by the heterozygous individuals, in terms of resistance to malaria, is bought at the price of a markedly reduced life expectancy for their homozygous siblings.

So far the terms "variants" and "mutations" have been used almost interchangeably—is there a difference? The answer is that there is no fundamental difference. The word "variant" is normally used to describe a sequence element present in the genome of 1% or more of the population. Such sequences are readily detected because they have established themselves. Any variant, which is present in the population at a level of less than one in a hundred individuals, is referred to as a mutation. These may be sequences that have recently arisen, and not yet had the opportunity to be positively selected, or they may be sequences that are to some extent detrimental, and are in the process of being eliminated by natural selection.

1.7 Genetic Drift

Organisms accumulate random mutations in their genomes—a process termed genetic drift. This was first made clear in a classical experiment carried out by Luria and Delbrück, which showed that the acquisition of resistance of a bacterium to attack by a pathogenic virus arises by chance in a few individuals in a normal bacterial population, prior to challenge with the virus. If the virus was not around, then the mutation provided no fitness advantage, and so it would be rather quickly selected out. However, if the virus happened by, then those few individuals carrying a random mutation that conferred resistance survived, while all the others died [6]. The Luria and Delbrück experiment makes the enormously important point that when an evolutionary problem arises organisms do not set out deliberately to find a solution to it. Instead, the solution is already lurking in the population in the form of random mutations. This is very different from the way we deal with problems in our daily life. When we face a problem, we try to analyse it rationally, and then design the best possible solution. We move from the problem towards the solution. Evolution works the other way round. Potential solutions, present in the form of mutations in the genomes of some members of the population, are just waiting there for an appropriate problem to come along.

Eukaryotic animals tolerate the retention of mildly deleterious mutations much better than do prokaryotic forms for two reasons. First, in contrast to haploid pro-karyotes, eukaryotes are diploid for at least part of their life cycle, and hence have two different copies of each gene, one inherited from mother and the other from father. For the majority of autosomally encoded genes, destruction of one copy has little if any discernible effect because, for most genes, one copy is quite sufficient. Because of this, eukaryotic genomes have a much greater capacity to tolerate genetic drift than do the genomes of haploid organisms. The second reason is that natural selection is a race. When you run a race over a distance of say 200 m, then the question of whether you will be among the first three to cross the finishing line will depend partly on your sporting capacities—but it will also depend crucially on who your competitors are. Suppose that we line up the entire human race on the starting line, then the probability that you will be first, second or third across the finish line is going to be vanishingly small. If, however, we select randomly ten members of humanity for this race, then the situation becomes very different. I might be one of your competitors, and I assure you that I pale at the mere thought of running. Another competitor might be bursting with energy but, being only 5 months old, is not going to be any serious problem for you. In other words, by limiting the number of randomly selected starters in the race, your chances of winning a medal improve enormously. The point is that since natural selection is a race, its effective-ness is directly proportional to the size of the population. Since we are talking about germline transmission here, the term "population" refers to that fraction of the total population that is involved in reproduction. Bacterial or viral populations are truly vast in size, and each member of the population is attempting to reproduce. Natural selection here is therefore highly effective, and gene products will be optimised to carry out the specific function for which they have been selected. By the same token even mildly deleterious mutations, acquired by random genetic drift, will be quickly selected out in these huge populations where competition is fierce and only the fittest survive. In eukaryotes, in contrast, because of their small effective population sizes, natural selection is much less stringent and so mildly deleterious mutations can accumulate in the germline. Genetic drift in diploid eukaryotes thus provides for a much greater store of tolerated mutations than can be supported in most prokaryotes. As a result, we have a much larger number of potential solutions just waiting for the appropriate problem to come along.

A large repository of germline-encoded mutations accumulating in eukaryotes is more than just a theoretical possibility, for genomic sequencing has demonstrated the presence of large numbers of variants in human genes. At least for the coding regions there is, on average, one variant found in the human population every eight base pairs [5]. Indeed, each person reading this sentence will in all probability harbour, at the very least, 85 genes in which one copy has been destroyed by mutation, and 35 genes for which both copies have been destroyed. But this is not all. Susan Lindquist and her co-workers demonstrated in the fruit fly *Drosophila* that genetic drift has resulted in many genes having acquired mutations that result in protein products that, while still functional, are almost literally falling apart [7]. These gene products are far from optimised. Their structural deficiencies may make them less than perfect at doing the job for which they were selected, but by the same token, they may have a much better chance than an optimised structure to take on some new function. All such mutant forms can be regarded as preformed "potential solutions". Nowhere is this clearer than in the adaptive immune system (see Chap. 4) for here every single lymphocyte that is newly generated from stem cells in the bone marrow produces a novel mutant antigen receptor. The resulting vast number of clones of mutant receptors can be considered preformed solutions, which are available to detect any and all pathogens.

1.8   Recombination

Because of genetic drift, mutations accumulate all the time. Very few of these mutations are in any way beneficial, and many are frankly deleterious. This raises a problem. Since mutations are happening all the time, what is to be done with the constantly increasing load of deleterious mutations? For mutations that completely disrupt the expression of essential genes, the answer is clear. The organism carrying such a lethal mutation will simply die, and hence the mutation is removed from the population. However the more interesting—and more common phenomenon—is that of mutations which are mildly deleterious, but not actually lethal. Such deleterious mutations will inevitably slowly accumulate. This awful scenario, first pointed out by the American geneticist Hermann Muller, became known as Muller's ratchet. The basic idea is that, as time goes by, there will be a gradual, continuous, irreversible increase in the number of far from perfect genes in each individual. If nothing is done about it, then this will inevitably result in extinction. How is extinction to be avoided?

Muller's resolution of this problem was to suggest that sexual reproduction and its accompanying meiotic recombination provide a chance to remove deleterious mutations (Fig. 1.1). The figure illustrates this with a hypothetical organism with a single chromosome. Each individual has inherited one copy of the chromosome from its mother and one from its father. During meiosis random recombination allows for rearrangement of the information prior to the formation of the haploid eggs or sperm. Many of the rearrangements will make little difference, but occasionally a lucky set of recombination events will result in eggs or sperm that have a substantially reduced mutational load. This resolution of the Muller ratchet paradox is widely held to have been a major selective force driving the evolution of sex.

Recombination, however, does more than merely permit natural selection to coddle advantageous mutations and to eliminate detrimental ones. It generates new associations of old genes and, because genes are archetypal team players, this can have surprising results. Each human being can be regarded as being formed from a team of 20,000 integrated genes. Almost all of these genes exist as variants or "alleles" in the population and so, except for identical twins, none of these teams

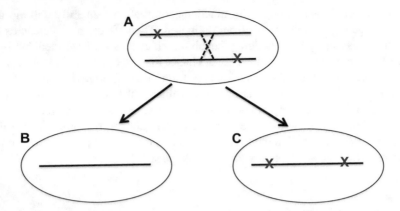

Fig. 1.1 Sex resolves Muller's ratchet. The heterozygous diploid individual A carries two deleterious mutations (x). Crossing over during meiosis 1 results in haploid gametes B and C. In this case gamete B now has neither deleterious mutation, while gamete C has acquired both

are identical. At every generation, recombination shakes up the composition of the teams to create new combinations of alleles of the 20,000 genes, and hence create unique new individuals.

In this sense the 20,000 genes of the human genome can be thought of as forming a genetic house of cards, so that removing or changing the shape or size of a card may cause reverberations throughout the whole structure. This bedevils the analysis of many genetically defined traits. One can see this in normal characteristics such as human height. In this case some 80% of the phenotype is heritable, and hence due to genetic factors. The rest is due to environmental factors such as nutrition. From large scale gene-mapping experiments it has been shown that around 10% of the variation in height is associated with the expression of some 180 genes, and that over a thousand genes contribute to the rest [8]. Trying to determine the height of an individual just by looking at his or her genes is therefore a pretty hopeless undertaking.

If one now looks at this in terms of human health, then it will be clear that a mutation that results in the loss or altered function of an essential protein may result in a genetic disease, that is to say, one that is heritable. There are well over 1500 known "monogenic" human diseases whose causes can be attributed to the disruption of a single gene [9]. However, many of these diseases are exceedingly rare. Much more common are diseases such as schizophrenia, rheumatoid arthritis or inflammatory bowel disease that, like height, are associated with the interactions of large numbers of genes. In these cases some of the "disease" alleles may be nothing more than otherwise perfectly normal variants that happen to find themselves in a less than suitable genomic team.

Recombination is thus a crucial player in evolution, and the French molecular biologist François Jacob expressed this point as follows: "Novelties come from previously unseen association of old material. To create is to recombine" [10].

1.9 Host–Pathogen Interactions Are an Arms Race

In this book we will be focussing on the fitness benefits of being able to survive attack by pathogens. It may be helpful, when considering human pathogens, to realise that there is nothing personal in what they are doing to us. All they want is a bit of peace and quiet, plus sufficient energy and resources to raise a brood of baby pathogens. The problem with them is that they have come to regard us as a source of energy and resources. We, in turn, do not take kindly to this and attempt to discourage them by getting involved in a form of a mutual arms race in which energy and resources have to be diverted from the centrally important task of generating progeny, to that of defence. Reciprocal arms races are commonplace in nature, for every competition between species for resources, every halfway success-ful marriage, and, in particular, every host–pathogen interaction follows this general scheme. The consequence is that while our immune system helps us to avoid or overcome infections, it also, and at the same time, exerts selective pressure on the pathogen, and thus drives the evolution of novel virulence mechanisms. These in turn exert selective pressure on the host to drive the evolution of the immune system. The result is an everlasting arms race between pathogen and host.

1.10 The "Generation Gap"

Pathogens accumulate novel virulence strategies in the form of useful mutations, which are then fixed in the population by natural selection. We, as the unfortunate target for their attack, must respond with appropriate resistance strategies, and these we too acquire by random mutation. From all of this one might think that, so long as we keep on doing our genetic homework, and counter the pathogen's new virulence strategies with ever newer complementary resistance strategies, we will be just fine. To a large extent this is true, though there is a terrible asymmetry built into the struggle. It is that for a mutation to be evolutionarily relevant it must, in general, be fixed in the germline, so that it can be passed on from one generation to the next. This means that, to a rough first approximation, the rate of fixation and hence the rate of evolution is measured not in hours, days or weeks but rather in terms of generations. However, many viruses and some bacteria have a generation time that is around 20 minutes while our generation time is, at best, 20 years. This means that there is an enormous "generation gap", for the difference between 20 minutes and 20 years is a factor of just over 525,000. Thus, such pathogens can evolve new, heritable, potential virulence strategies around a half a million times faster than we can respond with germline encoded potential resistance strategies. Given this massive evolution-ary advantage for the pathogens, one might well ask how on earth we could possibly have survived.

There are at least two answers to this question. The first applies to all eukaryotes, and it depends on the fact that eukaryotes are able to retain mutations introduced by genetic drift to a far larger extent than can prokaryotic organisms. Eukaryotes thus have a large pool of potential genetic solutions to future problems, and this may be, at least in part, what evens up the scales, which would otherwise be tipped so strongly in favour of pathogens with their short generation times.

A second, quite different, answer to the question of how we survive in the face of the enormous generation gap advantage for potential prokaryotic pathogens arose with adaptive immune systems. In adaptive immunity the generation gap is essentially abolished by the simple, but effective, trick of shifting selection from the germline to the soma. This process will be discussed in more detail in Chap. 4.

1.11   Evolution of Immunity

The first free-living eukaryotic animal forms were probably single celled amoebae-like organisms crawling around in shallow coastal waters eating any bacteria they might come across. Such tiny creatures would of course have represented an Eldorado of energy and metabolic resources, in what was doubtless otherwise a rather desolate environment, and hence they would have been the natural target for any viruses and bacteria in the neighbourhood. In order to survive, these amoebae must have quickly come up with effective means of defence. We should not underestimate the challenges facing these early unicellular eukaryotic organisms, for they were forced to deal with numerous quite different types of threat simultaneously. Transposons, viruses, bacteria and other predators would all have threatened their existence, and would have driven the selection of multiple defence strategies differing drastically in their mode of action. In immune defence any strategy—no matter how bizarre—that works, and thus leads to increased fitness, will be welcomed and adopted. One example of this can be seen in the means used by some free-living amoebae to protect their populations from destruction by viruses. Amoebae are attacked by giant Mimiviruses that establish a so-called virus factory in the amoeba's cytoplasm. Within this factory the Mimivirus assembles its progeny, after which the host amoeba cell is lysed. However, life is not a bed of roses for the Mimivirus, for they are themselves attacked by small Maviruses that sneak into the Mimivirus "factory" in the amoeba's cytosol, and hi-jack the resources concentrated there. In one further twist to the story some strains of amoebae, working on the principle "my enemy's enemy is my friend", have arranged to integrate the Mavirus into their own genome. When a Mimivirus attacks such an amoeba, the integrated Mavirus is activated and starts to replicate so that both Mimivirus and Mavirus particles are formed. The infected amoeba cell then dies, but the large number of Mavirus particles produced, reduces the effectiveness of Mimivirus infection of neighbouring amoebae by two to three orders of magnitude

[11]. In this way the amoeba has recruited a natural enemy of the Mimivirus to protect its own population.

The willingness to try anything that would help, even to a small extent was, perhaps, what spurred the initial diversity of the immune system, and set the stage for the numerous variations that came later. An amoeba, what ever else it may be, is first and foremost a crawling immune system. Of course we don't know how immune defence developed in molecular terms, but we have a pretty clear idea of what the general outline was, for the history of immunity is one long set of panic reactions to ever more sophisticated pathogen virulence strategies. Any new resistance mechanism that helped, even to a small degree, and thus increased host fitness would be honed and perfected by random mutation and natural selection, and become incorporated into the immune armoury. The fact that genes involved in immune defence are among the fastest evolving elements in any genome shows that this is what happened time and again during the course of evolution—and that the process has no end.

The immune system is an unparalleled example of the often unexpected ways in which the optimal compromise between investment of resources and fitness are sought. However, immunity's seemingly simple primary role of providing us with defence against pathogens is hedged about with innumerable ifs and buts. The initial stages of a pathogen attack are no time for molecular musings or second thoughts—what is required is a hard-wired constitutive innate response that can be applied right away. Yet, on the other hand, pathogens may well evolve ways around any given defence mechanism. Because of this, these innate defence mechanisms should ideally be redundant, diverse and based on alternative molecular categories. For such a system to function with the necessary speed, not only is a hard-wired response needed, but some form of memory will also be an important component, for the ability to recall the nature of a threat that has once been faced and overcome will make it easier to deal with it a second time round. In addition, it is crucial that the defence system be not only rapid, but also appropriate. It must effectively neutralise pathogens yet avoid attacking symbionts, commensals and "self" tissue. It must manage to do all this using the minimum number of genes, and yet give each individual in a species the best possible chance of survival under often rapidly changing environmental conditions. Particularly for animals whose generation times are long in comparison to those of their pathogens, reliance on the acquisition of suitable resistance mechanisms in the germline becomes a dangerous gamble, and immunity has evolved the means to hedge that bet by exploiting somatic as well as germline mechanisms of adaptation.

Maximising fitness involves maximising the number of progeny that survive and themselves produce fertile offspring. Clearly, any activity that drains resources from this essential business of reproduction must inevitably lead to a reduction in fitness. In this context, immunity seems to be an astonishingly wasteful system that squanders energy and resources as if there was no tomorrow. For example, in a human being thousands of millions of immune cells such as granulocytes and lymphocytes are produced every single day, only to be almost immediately destroyed—unused. This involves a massive investment of energy and resources, so much so that one often ends up with the thought that there must be a better way of doing things. That, however, is not a useful standpoint. Immunity is an evolutionarily stable strategy because the fitness benefits it provides more than counter the fitness loss caused by the apparent squandering of resources. It is this balance between resource expenditure and fitness, which is achieved by the sometimes bizarre-seeming mechanisms of immunity, that we will be looking at in the rest of this book.

References

1. Mayr E (1961) Cause and effect in biology. Science 134(3489):1501–1506
2. Hodgkin J, Barnes TM (1991) More is not better: brood size and population growth in a self-fertilizing nematode. Proc Biol Sci 246(1315):19–24
3. Doolittle WF (2013) Is junk DNA bunk? A critique of ENCODE. Proc Natl Acad Sci U S A 110 (14):5294–5300
4. Hsu E, Du Pasquier L (2015) Pathogen-host interactions: antigenic variation v. somatic adaptations. In: Richter DJ, Tiedge H (eds) Results and problems in cell differentiation, vol 57. Springer, Heidelberg
5. Lek M et al (2016) Analysis of protein-coding genetic variation in 60,706 humans. Nature 536 (7616):285–291
6. Luria SE, Delbruck M (1943) Mutations of bacteria from virus sensitivity to virus resistance. Genetics 28(6):491–511
7. Rutherford SL, Lindquist S (1998) Hsp90 as a capacitor for morphological evolution. Nature 396(6709):336
8. Lango Allen H et al (2010) Hundreds of variants clustered in genomic loci and biological pathways affect human height. Nature 467(7317):832–838
9. OMIM (2019) Online Mendelian inheritance in man, OMIM®. McKusick-Nathans Institute of Genetic Medicine, Johns Hopkins University, Baltimore, MD
10. Jacob F (1977) Evolution and tinkering. Science 196(4295):1161–1166
11. Fischer MG, Hackl T (2016) Host genome integration and giant virus-induced reactivation of the virophage mavirus. Nature 540(7632):288–291

Further Reading

Darwin C (1859) On the origin of species. John Murray, London
Howard J (1982) DARWIN a very short introduction. Oxford University Press. isbn:0-19-285454-2
Jacob F (1977) Evolution and tinkering. Science 196(4295):1161–1166
Malthus T (1798) An essay on the principle of population. www.esp.org/books/malthus/population/malthus.pdf

Chapter 2
From Unicellular to Metazoan Immunity

In a population of single cell organisms, each individual must optimise its capacities for all of the essential tasks in life, such as catching food, moving from place to place, reproduction, defence and so on. These different components of fitness compete with each other for energy and metabolites, so that trade-offs become inevitable. With so many divergent goals to be met, it is clear that there will be no perfect solution to each of the organism's many problems. Yet, because only the fittest will survive, the pressure to find the best possible compromise solutions is considerable, and those that have evolved are very good indeed. This is what makes unicellular life such a very successful evolutionary strategy.

If a multicellular life form is to evolve from a population of unicellular animals, then it must satisfy the minimal criterion that its overall fitness be at least equal to the average fitness of the members of the single cell population. To achieve this, the way forward for the multicellular organism is to assign the various fitness components to different groups of cells. One group is responsible for movement, another for reproduction, another for catching food and so on. Because each group of cells specialises on a particular set of functions, it is freed from the constraints that now apply only to other groups, and so the requirement for making trade-offs is reduced. In this way, the overall fitness of the multicellular organism can be increased beyond the level achieved by its unicellular rivals. The price, however, is that the fitness of the individual cells falls to zero—a lymphocyte, for example, has no chance of surviving as an independent life form in the big wide world outside, for far too many of the skills that it would require are now the specialised competences of other cells. The problems of adapting to a metazoan life should not be over emphasised, for many unicellular animals cycle through various different cell forms during the course of their lives. One can think of them having a number of different cell types "one after another" instead of "all at once" as in metazoans [1]. Nevertheless, the upheavals involved in the transition from unicellular to metazoan life are considerable.

We will look at four situations that impact on the architecture of immune defence in metazoans, where this transition has forced changes in what were tried and tested

© Springer Nature Switzerland AG 2019
R. Jack, L. Du Pasquier, *Evolutionary Concepts in Immunology*,
https://doi.org/10.1007/978-3-030-18667-8_2

solutions to some of unicellular life's problems. The first of these is the distinction between germline and soma. The second involves the restriction of cellular competences as exemplified by the capacity to carry out phagocytosis. The third concerns the breakdown of the clear distinction between self and non-self, which metazoan life brought with it. The fourth concerns the requirement for a completely new function—rapid, long-range mobility of immune cells, and the cellular communication system that this requires.

2.1   Germline and Soma

The processes of replication and transcription both require that the double-stranded genomic DNA be locally melted so as to serve as a template. Anything that involves tampering with the structure of the genome is potentially mutagenic, and because of this, replication and transcription are genetically hazardous activities. This problem is especially acute in the protozoan ciliates, whose large cells require constant gene transcription to provide for the production of the huge amounts of proteins that they continuously need. They solve this problem by duplicating the zygote nucleus. One nucleus is retained as a germline "micronucleus", which is used only in the formation of the next generation. The second nucleus, in contrast, serves the somatic demands of the cell by turning into a "macronucleus". The details of what happens during this transformation differ considerably from species to species, but the common thread is that the chromosomes in this somatic nucleus are replicated several times, and most of the non-coding genomic DNA is destroyed. This leaves a macronucleus containing multiple copies of the coding regions of genes but lacking most of the other regions of the genome. The macronucleus has thus lost a great deal of the genetic information, and hence cannot serve in sexual reproduction, but on the other hand, amplification of the genes required for daily transcription ensures a continuously high rate of protein synthesis.

In metazoans, the dangers of overuse of the nucleus are also present, but the division of labour, characteristic of metazoan life, allows the germline genome to be relatively protected by being sequestered in a separate cell lineage formed very early in development. This germline lineage in metazoans provides a protected source of genomic information reserved for reproduction. In this sense it can be thought of as being the metazoan equivalent of the ciliates' micronucleus. All the other somatic lineages are there to provide for the life of the current generation and so require less stringent protection. They can be thought of as the metazoan equivalent of the ciliates' macronucleus. It is the separation of germline and soma that makes protein-based somatically diversified immune systems possible in vertebrates, for the construction of such a system requires that the DNA of lymphocytes be subjected, for a brief period, to a very high rate of recombination and mutation—a treatment that would be non-sustainable if applied to germline cells.

2.2 Phagocytosis: An Old Habit Becomes Restricted

One consequence of the division of labour between different cell lineages is that functions essential in lineage A may be irrelevant, or frankly deleterious, if expressed in lineage B. One simple example of this sort of compartmentalisation of function is seen in the ability of cells to take in material from the outside world by phagocytosis. This is a general competence in unicellular predators, for free-living amoebae must constantly search for food, and then internalise it by phagocytosis. Both the searching and the internalisation are activities that are dependent on the cytoskeleton—a highly dynamic structure, composed of a plethora of different structural proteins and regulatory enzymes. The cytoskeleton gives the cell its shape, and enables both the movement of organelles within the cytosol, as well as the movement of the cell with respect to the substrate. During phagocytosis, the local structure of the cytoskeleton is altered so that the plasmalemma flows around the target particle, and encloses it in a membrane-bounded vacuole—the phagosome. Within the cytosol, the phagosome goes through a process of maturation, during the course of which it may be fused with lysosomes, or other preformed storage vesicles packed with hydrolytic enzymes. Activation of these enzymes within the resulting phago-lysosome causes the contents of the vacuole to be degraded into simple metabolites, which are then available to fuel the cell's metabolism.

2.2.1 Lineage Restriction of Phagocytic Competence in Metazoans

Phagocytosis is a very basic, and hence ancient, competence of unicellular eukaryotic animal cells. It is, however, an activity that must be restrained in the cells of metazoans, for an organism, whose cells are all constantly trying to eat their neighbours, is unlikely to be a winner in the struggle for existence. The requirement to tone down phagocytic activity in multicellular animals can be seen, at the very simplest level of multicellular complexity, in the social amoeba *Dictyostelium discoideum*, which lives much of its life as a unicellular organism crawling around feeding on bacteria by phagocytosis. It is, however, also conditionally a multicellular organism for, once the available bacteria have all been eaten, the single cells crowd together and form a so-called slug, which may contain up to around 100,000 individual cells. The slug can move around, searching for a fresh supply of food, but if none is found, it will attach to the substrate, form a stalk with a fruiting body on top and generate spores, which can then be wafted away to pastures new.

The interesting part of this story, from our point of view, is that, as the slug is formed, the individual amoebic cells sacrifice some of their competences in order to contribute to the life of the new community. One of the first things to go is the amoebic form's enormous capacity for phagocytosis. However, since the slug is a rich wandering source of energy and metabolites, it must be able to protect itself from attack by pathogens, and so one of the first specialisations to take place is that around 1% of the cells retain their phagocytic capacity, and are charged with the business of defence. These so-called sentinels patrol through the mass of cells, picking up bacteria and debris and dumping them outside of the slug [2]. In this situation phagocytosis has been switched from being an essential skill of every cell, to being a competence reserved for a small population of specialised cells.

In more complex animals there is a similar restriction of highly active phagocytosis to certain specific cell types, such as macrophages or granulocytes, though many other cells do retain it as a reserve competence.

2.2.2 Professional Phagocytes Exploited by Pathogens

Quite apart from the need to discourage phagocytic cannibalism, there is a further advantage in restricting phagocytosis to a small number of specialised lineages, and that is that a cell which is able to engulf particles, runs the risk that pathogens will target this activity, exploiting it to gain entry to the cell. True, the maturation of the phagosome, involving vacuole acidification, activation of proteases, and the activation of an oxidase system in the phagosome membrane that allows for the formation of bactericidal reactive oxygen species inside the vacuole, do all combine to make life inside a phagosome challenging for a bacterium. And yet, despite all that, neutralising potential pathogens by eating and digesting them is not a fool-proof means of defence. Host–pathogen interactions follow the concept of an arms race, and in consequence, some pathogens have come up with effective counter-moves. One that has done so is *Legionella pneumophila*, which infects a range of freshwater amoebae. The strange career of this bacterium as an agent of interest in human medicine not only illustrates the ways in which phagocytosis can be subverted by pathogens, but also shows the remarkable conservation of the biochemical details of the phagocytic pathway over long periods of evolutionary time.

A range of freshwater amoebae phagocytose *L. pneumophila* as if it were food, but once safely inside the cell, the pathogen modifies the phagosome's ability to interact with the cytoskeleton. To do this the bacterium inside the phagocytic vacuole builds a syringe-like structure with which it injects some 300 different bacterial mediators through the vacuole membrane into the cytosol. These mediators subvert the activities of the cell's cytoskeletal machinery and, by doing so, prevent fusion of the phagosome with lysosomes [3]. Having thus escaped certain destruction, the bacterium replicates within the modified vacuole, and its progeny are then liberated from the infected cell. This pathogen has thus evolved the means to exploit features of an ancient mechanism of cell biology—the formation and maturation of phagosomes—to enable it to infect its host. Mammalian phagocytes make use of the same ancient mechanisms for phagosome formation and processing as are used by the amoebae, so that the lessons *L. pneumophila* learned by infecting amoebae do, under the right circumstances, give it the opportunity to turn into an "accidental pathogen" of humans. This is what happened in the outbreak of Legionaires' disease in Philadelphia in 1976. As far as is known *L. pneumophila* was never a significant human pathogen prior to man's invention of air conditioning systems that require large tanks full of cooling water. These freshwater tanks provide a new ecological niche for amoebae, and hence also for the pathogens that prey on them. The air conditioning units, fed from these tanks, generate aerosols and, if the water they use is contaminated, then the aerosol will contain suspensions of *L. pneumophila*. Once inhaled, the bacterium infects alveolar macrophages in the lung and then diverts the phagosome from its normal maturation route, in a way that closely mirrors the events taking place during infection of the amoebae. *L. pneumophila* has thus evolved a virulence strategy, which targets basic aspects of phagocyte cell biology that have remained unchanged since the last common ancestor of humans and amoebae over 600 million years ago.

2.2.3 Pathogens Can Reactivate and Exploit Phagocytic Competence

Of course the basic processes of cytoskeletal structure and dynamics, on which phagocytosis is based, are still present in non-phagocytic cell populations of complex metazoans, and they can, under certain circumstances, be reactivated. For example, species of *Salmonella, Shigella, Bordetella, Listeria, Chlamydia* and others, all make use of a virulence mechanism by means of which they can induce certain non-phagocytic mammalian cells to ingest them. Like *Legionella,* these bacteria too build syringe like structures to inject bacterial mediators into the cell's cytosol, where they take control of the enzymes, which locally regulate the growth of the cytoskeleton [4]. By doing so they force the membrane to extend around, enclose and then internalise the pathogen. These pathogens have also evolved various different strategies, which empower them either to hinder the fusion of the vacuole with lysosomes, or to escape from the phagosome into the cytosol. They thus all exploit the latent phagocytic apparatus in these cells to gain a safe haven within which they can replicate.

2.2.4 Diversification of Phagocytic Competences in Metazoans

This initial restriction and specialisation of phagocytic competent cells, seen in *Dictyostelium discoideum,* are driven by the eternal struggle between host and pathogen to ever-higher degrees of sophistication, for as the complexity of multicellular life increases, so too does the number of different tissues and niches in the body. The threats that potential pathogens pose to the mucosal surface of the lung are very different from the threats likely to be encountered in the gut mucosa.

These, in turn, are of a completely different order from the problems arising from an infection in skin, heart muscle or liver. Because of this, a simple generalised phagocytic cell type, like *Dictyostelium discoideum*'s sentinel cells, soon becomes inadequate to deal with all of the threats. At the same time, as in all host–pathogen interactions, the nature of the threat is continually changing as improvements in the defence system drive the selection of new virulence strategies in the pathogens. This in turn drives selection of ever better defence system architectures, and results in the multiplication of specialised cell populations in the immune system. This speciali-sation, and consequent divergence of different populations, is already evident in many different invertebrates, while in mice and humans, the number of identified distinct populations of cells with phagocytic capabilities increases continuously as analysis techniques improve—and currently no end to this is in sight.

2.2.5   Autophagy: A Eukaryotic Cousin of Phagocytosis

Eukaryotic cells are equipped with a means of engulfing particles in their own cytosol by a process of "internal phagocytosis" known as autophagy. Autophagy plays several important roles in the cell, including that of serving as a means of ensuring intracellular hygiene. Eukaryotic cells dispose of damaged or denatured proteins by shuttling them into a cell organelle called the "proteasome" where they are digested by proteases. However, the proteasome has a fairly narrow entrance. Aggregates of denatured proteins may soon grow to a size too large to enter, and in that case they accumulate in the cytosol. In a long-lived or metabolically active cell the danger is that the cytosol gradually fills up with this sort of aggregated garbage. The solution has been that such unwanted rubbish is disposed of by being enclosed within a membrane bound vacuole, which is then fused to lysosomes. Lack of nutrients also switches on this autophagy process so that the cell can tide itself over a period of famine by engulfing and digesting its own contents—an ageing and damaged mitochondrion here, perhaps a piece of endoplasmic reticulum there. The cell can also use this system to engulf and dispose of cytosolic pathogens [5].

Removal of cytosolic bacteria by autophagy brings with it a problem of self-non-self discrimination because mitochondria have evolved from bacterial endo-symbi-onts, and they retain to this day many prokaryotic features. Nevertheless, while healthy mitochondria should be left in peace, invading bacteria must be quickly located and destroyed. One well-established way of achieving this involves an indirect means of detecting pathogens escaping from the phagosome vacuole into the cytosol. This makes use of the fact that the inner surface—but only the inner surface—of the phagosome membrane is decorated with glycans rich in galactose. Being inside the phagosome, these glycans are "invisible" to galactan receptors—the galectins. However, if the pathogen tries to escape from the phagosome, then it must first damage or destroy the phagosome membrane, and by doing so it exposes these inner membrane glycans. As soon as that happens, the cytosolic galectins rapidly bind the galactose rich inner membrane glycans. These glycan–galectin complexes then act as "eat me" signals for the autophagy machinery [6].

2.3   Metazoans Are Societies of Cells: The Importance of "Self"

The transition from unicellular to multicellular forms led, quite literally, to a "whole new way of life" dominated by the need to adjust to the special needs of a society of cells. The fitness of a metazoan organism depends crucially on the cooperation of all the different cells in its body, and this in turn requires a large degree of tissue organisation and integration. For this, cells must have a sophisticated means of knowing who they are, where they are and with whom they should be interacting. They also need to know the current state of play as it affects the tasks they have been allotted to deal with. They must, in other words, have a considerable ability to access and interpret information about "self". For these purposes, multicellular organisms have evolved a whole battery of surface molecules with which their cells can recognise and identify their neighbours, and they have also evolved receptors for soluble mediators, like hormones or cytokines, which bring information as to what problems are being faced in more distant parts of the body. Thus, while the individual cell in a population of single cell animals is primarily concerned with information about the outside world of "non-self", the individual cell within a multicellular organism is primarily concerned with information about "self". Recognition of "self" is thus an essential part of the normal homeostatic regulation of the society of cells which constitute a metazoan organism.

Cells of the immune system, like all other cells in the body, read this sort of "self" information, and it is crucial for the functioning of immunity in response to a pathogen attack. However, a mobile defence system equipped with the ability to engulf particles by phagocytosis has enormous potential beyond providing defence against pathogens. Over the course of evolution, immunity has taken on a number of other jobs, which require that it can deal not only with pathogens, but also with several different forms of "altered self". Among these are "apoptotic-self", "necrotic-self" and "oncogenic-self".

2.3.1   Recognition of "Apoptotic-Self"

When cells are, for whatever reason, no longer required, they may be directed into a "default differentiation pathway" that results in their death by apoptosis. Apoptosis is by no means restricted only to excess cells generated during embryogenesis, for in an adult human being countless billions of cells are driven into apoptosis every single day. The major problem with apoptosis is that somebody has to clear away the corpses. In the nematode worm, which has no dedicated immune system, an apoptotic cell is removed by a neighbouring normal cell, which briefly activates a phagocytic program, and swallows the corpse. In more complex organisms such as mammals, epithelial cells, endothelial cells and fibroblasts may likewise operate as "amateur" phagocytes in certain circumstances to remove corpses. However, it is the "professional" phagocytic cells of innate immunity—the macrophages—that carry out the major part of apoptotic cell clearance. The dying target cell releases low molecular weight diffusible "find me" signals that act as attractants, and it also expresses a separate set of membrane bound "eat me" signals, such as phosphatidylserine, which together attract the attention of macrophages. These signals result in activation of the phagocyte, which will then engulf the apoptotic cell. The phagocyte's engulfment program and the apoptotic program expressed in the target cell have co-evolved, and this is of critical importance, for it ensures that the phagocytes do not engulf normal cells and hence cause random tissue destruction. This engulfment process is normally very rapid, so much so, that even in organs in which there is a massive rate of apoptosis, as in vertebrate bone marrow or thymus, dead cells can rarely be seen [7].

2.3.2 Recognition of "Necrotic-Self"

Another form of cell death is referred to as necrosis. Here, however, we enter a grey area where the distinction between "self" and "non-self" becomes blurred, for recognition of necrotic-self, unlike recognition of normal-self or apoptotic-self, has consequences in terms of further activation of the immune system. The crucial phenomenological difference between necrosis and apoptosis is that in necrosis the integrity of the cell membrane is not maintained, and cytosolic components, which are normally never encountered outside of cells, are released.

Necrotic cell death can result from a number of quite different causes. It occurs at sites where the rate of apoptotic cell clearance is inadequate, at sites of sterile traumatic injury, or where virus infected cells rupture and die. The latter case is one in which it is vital that the immune system be alerted, and the means which have evolved to do this is that the phagocytes interpret some of the cytosolic components released from necrotic cells as "danger" signals. The response to necrotic cell death is correspondingly different from the response to the "eat me" signals on apoptotic cells. In both cases the cell debris is removed by phagocytosis, but whereas the removal of apoptotic corpses by phagocytes is a process of "silent clearance" that is carried out without further activation of the immune system, the clearance of necrotic material may induce an inflammatory reaction leading to an immune response.

2.3.3 When Cooperation Is Not Enough: Recognition of "Oncogenic-Self"

In addition to being able to recognise and distinguish normal self from apoptotic-self and necrotic-self, immunity must, above all else, be able to detect pathogens. The means by which this is achieved will be the subject of the next two chapters, but it is worth mentioning here one final form of "self", which may be thought of as "pathogenic-self" or more precisely "oncogenic-self". All cooperative social systems are dependent on the individual members being prepared to make costly sacrifices for the good of the whole. In game theory this is referred to as "altruistic cooperation". However, cooperation on its own is never sufficient, for some individuals will quickly attempt to increase their personal fitness by taking more out of the system than they contribute. One simple example can be seen in bacteria that form mats. Each individual bacterium growing in a medium would ideally have both maximal access to the medium and maximal access to oxygen. Certain species achieve this by clubbing together to form a buoyant slime that holds all the members of the population at the liquid air interface. Everybody contributes to forming the buoyant slime, and everybody benefits from a place at this magic interface. Everybody is happy—for a while. However, some individuals inevitably start to cheat. They hang on to the advantages of life in the mat, but they cease to contribute to the production of the mat's buoyant slime. By doing so they save energy and resources, which can now be devoted to reproduction, and in this way they increase their fitness relative to that of their honest neighbours. They therefore constitute an ever-increasing fraction of the population and, inevitably, the mat soon collapses. Everybody loses [8]. Both theoretical analysis and observations in the field indicate that to hold such a social system together altruistic cooperation of the honest members must be reinforced with altruistic (that is to say costly) punishment, directed at the cheats [9]. In vertebrates the immune system is responsible for punishing the cheats.

The life of a human being is based on the collaboration of billions of cells each of which contributes to the survival of the organism as a whole, and in return each of these cells is supported with oxygen, metabolites and protection from pathogens. However, cells, which have undergone oncogenic transformation, have discovered, by random mutation, a means to break all the rules, and hence extract more than their fair share from the community. By doing so, tumour cells turn into endogenous pathogens. If the organism is to survive, these tumour cells have to be identified and destroyed—no matter what the costs may be. The immune system is charged with controlling these cheats.

2.4 Immobile "Epithelial" Immunity in Basal Invertebrates

At the simplest level of metazoan organisation—as in sponges—the embryo develops into an adult that lacks complex organs. The next step in increasing organisational complexity is seen in the Cnidaria where the embryo undergoes gastrulation to generate the two "germ layers" that will give rise to ectoderm and endoderm. Because these animals are composed of the products of these two layers, they are often referred to as "diploblasts". The final step in this progression involved a change in gastrulation such that three, rather than two, germ layers are established—and these give rise to ectoderm, endoderm and mesoderm [10]. These "triploblasts" are bilaterally symmetrical and are referred to as "the Bilateria" (Appendix A).

In diploblastic organisms, like sponges and Cnidaria, epithelial cells are responsible for immune defence. These cells are equipped with receptors able to recognise potential pathogens, and they respond to attack both by producing antimicrobial peptides, and by destroying invaders by phagocytosis. However, few of their pathogen-recognising receptors have been characterised in any detail so far. True, there are genome sequences available for comb jellies and sponges, for the placazoan *Trichoplax adhaerens,* and for a number of Cnidaria, and in these sequences, genes for molecules with similarities to quite a number of innate immune system receptors have been found. However, the main problem in interpreting such data is that one of the evolution's central mechanisms is to modify genes selected for one function, altering them so that they can do something new and different—a process known variously as "bricolage", "tinkering", "borrowing" or "exaptation".

A good example of this is seen with the so-called Toll Like Receptor (TLR) family of molecules (see Sect. 3.2.1.1). These TLRs are membrane-spanning proteins with a ligand binding domain on one side, and a domain that links them to the cytoplasmic signal transduction machinery on the other (see Fig. 3.2). In mammals there are usually 10 genes coding for TLRs, and the proteins they code for are all innate immune receptors involved in detecting various microbial ligands. In the fruit fly *Drosophila melanogaster* things are different. Here there are nine TLR genes, but "tinkering" has given them quite different functions. One of them does indeed code for an innate immune receptor that binds a viral protein [11], but another has been converted into a cytokine receptor [12]—and most of the rest are cell adhesion molecules [13]. In the nematode worm, *Caenorhabditis elegans,* tinkering has taken yet another direction. In this animal there is just one TLR. It is not an innate system receptor, but instead its expression is necessary for the development of certain sensory neurones [14]. Nevertheless, despite the fact that relatively little work has been done on the functioning of immunity in lower metazoans, there are a few bright spots, and both in the sponge *Suberites domuncula* [15] and in the diploblast *Hydra* [16] innate immune receptors have been identified and functionally characterised. In both cases the receptors are structurally reminiscent of bona fide innate receptors present in vertebrates.

The work on *Hydra* has demonstrated that even at this "lowly" level, innate immunity faces a much more complex task than just detecting and destroying any and all microbes, for *Hydra* possesses a rich collection of microbial symbionts. It has a bacterial "microbiome", whose loss results in defects in growth, and in addition some species form stable associations with certain green algae [17]. Thus even at the level of relatively simple diploblasts, this "epithelial" form of immune system is able to distinguish the category of "self plus friendly commensals" from that of "foes".

This sort of epithelial immunity is also relied on as the sole means of defence by some triploblasts, like the nematode worm *Caenorhabditis elegans*, but as animals became larger and more complex so the cells of the immune system became increasingly specialised and this required the evolution of ever more complex mesoderm-derived haematopoietic stem cell lineages to generate the various types of immune cells. By becoming more highly specialised, immunity had to evolve the means to assign resources wherever they might be needed and this required that cells of the immune system be mobile.

2.5   Immune Defence in Complex Metazoans Must Be Mobile

Mobility has to be a central element in the architecture of the defence system, since apoptosis, necrosis, tumour induction or infection can take place anywhere within a multicellular organism. This does not mean that all immune cells must necessarily be mobile, but it does mean that immunity has to be able to concentrate resources where they are needed. How is this organised? Defending a complex multicellular organism from attack by pathogens has something in common with the old problem of how best to defend a medieval town. For the sake of economy just a handful of sentries were left up on the wall to raise the alarm, and summon help in case of attack. This way of splitting defence into two components—detection and summoning—is very much the way that the immune system organises defence. Resident sentinel cells in the tissues—macrophages and dendritic cells—are principally there to detect the presence of "trouble", and then to send for help.

This defence architecture requires that the sentinel cells and the effector cells be linked through a communication system, and that the summoned effector cells be capable of precisely directed motion. It is in the nature of host–pathogen interactions that both the immune communication network and the mechanisms of directed effector cell movement become the targets of pathogen virulence mechanisms.

Movement of cells of the immune system through tissue depends on the same coordinated cytoskeletal activity that makes it possible for unicellular amoebae-like animals to get around. Early in metazoan evolution, cell movement was coupled with the ability to recognise "self" information expressed on the surfaces of other cells, so as to permit directed movement of a cell to a predetermined position within the organism. The immune system adopted these mechanisms to ensure that the right cells end up cooperating with the right partners at the right time, and in the right place. This, however, is, on its own, not enough. Immunity is a dispersed system, and in an adult mammal many of its cells have to be able to travel vast distances—vast when measured in terms of cell diameters—in the shortest possible time. Of the numerous novelties that evolved during the transition from invertebrate to vertebrate life [18], one of the most important—from the point of view of immunity—involved a change in the construction of the blood vessels. In invertebrates the vessels of closed circulatory systems are tubes made of extracellular matrix. In vertebrates these tubes are lined with a layer of endothelial cells. This converts the inanimate vessel wall of the invertebrates into an "intelligent" surface that makes possible the recruitment of precisely defined cell populations into and out of the circulation (Sect. 3.9.1). To exploit this option the immune cells must recognise the appropriate entry and exit sites of the blood and lymph vessels and, once again, this requires the ability to recognise the appropriate members of the suite of cell surface markers of "self", which are expressed on the vessels' endothelial cell surfaces.

2.6   Two Arms of Immunity: "Innate" and "Adaptive"

All animals use a so-called innate immune system, which will be discussed in more detail in Chap. 3. It is based on a broad repertoire of structurally diverse pathogen-sensing molecules linked to an equally diverse array of effector functions, the elements of which are all encoded in the germline. There are, in addition, systems of "adaptive" immunity that use alternative strategies to generate potentially limit-less receptor repertoires by means of somatic adaptations. These "adaptive" reper-toires will be considered in Chap. 4. But before coming to the nature of the innate and adaptive repertoires, it is worth looking briefly at the history of the discovery of these two arms of immunity.

The study of the mechanisms of immune defence really got under way when, in 1908, the Nobel Prize in Physiology or Medicine was awarded jointly to Paul Ehrlich and Ilya Metchnikoff. These two scientists had very different interests. Metchnikoff showed that particular cells in starfish larvae had the ability to engulf particles. This was the start of the study of phagocytosis, which lies at the heart of what is now known as innate immunity. Ehrlich, on the other hand, had struggled with the problem that the mammalian immune system was able to react specifically with many different things. In looking for an explanation for this remarkable diversity, Ehrlich came up with a model in which cells of the immune system were equipped with multiple different receptor molecules on the surface. Each of the different receptors was viewed as being able to bind a different ligand, and when one of these receptors bound its ligand then the cell responded by producing large amounts of this particular receptor. This proposal became the "leitmotiv" of immu-nology, and it evolved in response to advances in our understanding over the course of the next century.

2.6.1 The "Adaptive Problem" in Vertebrates: How Many Specificities?

Perhaps the first major upheaval in this regard was the work, centred on the ability to raise specific antibodies to a wide variety of simple chemical groups, and to carbohydrates, lipids or denatured proteins [19]. In 1936 this culminated in the publication of Karl Landsteiner's landmark book entitled "The Specificity of Serological Reactions". From this point on it seemed that the adaptive immune system could detect and make antibodies to absolutely everything. This raised two problems for Ehrlich's model. The first of these was that if adaptive immunity could recognise everything, then the immune cells must express an impossibly large number of receptors on their surface. How was this to be done, and how was the cell to determine which of these receptors had bound its ligand? How was the cell then to select just this one receptor and produce it in huge amounts as antibody? The Australian immunologist Burnet proposed a simplifying modification [20]. Instead of there being lots of receptors on every lymphocyte, there would be lots of lymphocytes each of which expressed just one receptor specificity. Recognition of a particular antigen thus resulted in the activation of only those cells, which expressed the appropriate receptor [21]. The second problem concerned the question of how receptors specific for so many different structures could be formed. On the one side were those who emphasised that the receptors must be coded, like every other protein, in a hard-wired fashion in the germline. On the other side, were those who felt that the number of different receptors which would be required to explain the enormous range and specificity of adaptive immune responses, would have to be so large that no genome could code for them all in the germline. Those favouring this second view therefore envisioned some sort of novel genetic mechanism, which would permit the elaboration of a vast repertoire of receptors in special somatic cells—the lymphocytes [22]. And so started the great "germline versus somatic generation of diversity" debate, and much of the interest and excitement in immunology went into devising experimental approaches, which would settle this question. This effort by many people culminated in the experiments, carried out by Susumu Tonegawa and his colleagues at the Basel Institute of Immunology, which finally provided the solution [23]. It turned out that though Ehrlich had provided the crucial conceptual framework, his explanation had been off the mark in almost all of its details [24]. Nevertheless, the long search for the basis of the diversity of vertebrate adaptive immune receptors made the twentieth century very much the age of Ehrlich and of "adaptive immunity".

2.6.2 Metchnikoff's Legacy

All this time the discovery of defence by phagocytosis eked out, at best, a shadow existence. Phagocytosis was all very well, but by comparison with the intellectual challenge of the generation of the diversity of lymphocyte receptors, it was very small beer indeed. Macrophages and other phagocytic cells seemed to mount a mere knee jerk reaction to the presence of foreign bodies, and so this arm of immunity was regarded as a relatively uninteresting, hard-wired, inborn or "innate" response. There was a widespread view that innate immunity is very ancient and ineffective while adaptive immunity is more modern and sophisticated. This view is entirely mistaken. It is, however, true that comprehensive, protein-based, adaptive systems are "modern" in the sense that they occur only in vertebrates. Perhaps, largely because it did not produce the sorts of headlines that guaranteed publication in all the best journals, innate immunity eked out a pauper's existence far from the lavish grants and glittering prizes available to those working on the adaptive system.

2.6.3 Phagocytosis and Janeway's "Dirty Little Secret"

A revolutionary turning point in the relationship of innate to adaptive immunity in vertebrates came with a truly seminal remark made by Charles Janeway, who pointed out that immunologists had what he referred to as a "dirty little secret" [25]. In reality the secret was one of those things that everybody was aware of, but nobody had paid much attention to it. As every immunologist knew, if one injected a pure protein into a mouse, then basically nothing happened. However, if one first mixed the protein with bacterial extracts, or other materials such as alum crystals, which stimulated phagocytic cells of the innate immune system, then the chances were good that the adaptive immune system would be activated and produce a splendid and highly specific antibody response. The conclusion was clear: the innate immune system was required to initiate a response of the adaptive system to a novel antigen. Janeway's unveiling of the "dirty little secret" ushered in the second age of immunology, an age in which Ehrlich's world is fused with that of Metchnikoff. This fusion is much more than just an empty intellectual exercise. Since innate and adaptive immunity have coexisted, and pursued common goals, for hundreds of millions of years, it is no real surprise that in this time frame these two rapidly evolving systems have been naturally selected, so as to optimise the common effort. They are now so tightly integrated that attempts to draw strict demarcation lines between them become futile.

Having said all this it is nevertheless clear that there is indeed one all-important area in which "innate" and "adaptive" systems are, and always will be, dramatically different. That crucial difference is that innate systems are always germline based, while adaptive systems are always based on somatic adaptations. These matters will be discussed in more depth in Chaps. 3 and 4.

References

1. Sebe-Pedros A, Degnan BM, Ruiz-Trillo I (2017) The origin of Metazoa: a unicellular perspective. Nat Rev Genet 18(8):498–512
2. Chen G, Zhuchenko O, Kuspa A (2007) Immune-like phagocyte activity in the social amoeba. Science 317(5838):678–681
3. Escoll P et al (2013) From amoeba to macrophages: exploring the molecular mechanisms of Legionella pneumophila infection in both hosts. Curr Top Microbiol Immunol 376:1–34
4. Puhar A, Sansonetti PJ (2014) Type III secretion system. Curr Biol 24(17):R784–R791
5. Zhang H, Puleston DJ, Simon AK (2016) Autophagy and immune senescence. Trends Mol Med 22(8):671–686
6. Randow F, Youle RJ (2014) Self and nonself: how autophagy targets mitochondria and bacteria. Cell Host Microbe 15(4):403–411
7. Brown GC, Neher JJ (2012) Eaten alive! Cell death by primary phagocytosis: 'phagoptosis'. Trends Biochem Sci 37(8):325–332
8. Rainey PB, Rainey K (2003) Evolution of cooperation and conflict in experimental bacterial populations. Nature 425(6953):72–74
9. Hauert C et al (2007) Via freedom to coercion: the emergence of costly punishment. Science 316(5833):1905–1907
10. Wolpert L (1992) Gastrulation and the evolution of development. Development 116 (Suppl):7–13
11. Nakamoto M et al (2012) Virus recognition by Toll-7 activates antiviral autophagy in Drosophila. Immunity 36(4):658–667
12. Valanne S, Wang JH, Ramet M (2011) The Drosophila Toll signaling pathway. J Immunol 186 (2):649–656
13. Pare AC et al (2014) A positional Toll receptor code directs convergent extension in Drosophila. Nature 515(7528):523–527
14. Brandt JP, Ringstad N (2015) Toll-like receptor signaling promotes development and function of sensory neurons required for a C. elegans pathogen-avoidance behavior. Curr Biol 25 (17):2228–2237
15. Wiens M et al (2005) Innate immune defense of the sponge Suberites domuncula against bacteria involves a MyD88-dependent signaling pathway. Induction of a perforin-like molecule. J Biol Chem 280(30):27949–27959
16. Bosch TC (2014) Rethinking the role of immunity: lessons from Hydra. Trends Immunol 35 (10):495–502
17. Rahat M, Dimentman C (1982) Cultivation of bacteria-free Hydra viridis: missing budding factor in nonsymbiotic hydra. Science 216(4541):67–68
18. Gee H (2018) Across the bridge: understanding the origin of the vertebrates. University of Chicago Press
19. Landsteiner K, van der Scheer J (1931) On the specificity of serological reactions with simple chemical compounds (inhibition reactions). J Exp Med 54(3):295–305
20. Burnet FM (1957) A modification of Jerne's theory of antibody production using the concept of clonal selection. Aust J Sci 20(3):67–69
21. Burnet FM (1959) The clonal selection theory of acquired immunity. Cambridge University Press, Cambridge
22. Brenner S, Milstein C (1966) Origin of antibody variation. Nature 211(5046):242–243
23. Hozumi N, Tonegawa S (1976) Evidence for somatic rearrangement of immunoglobulin genes coding for variable and constant regions. Proc Natl Acad Sci U S A 73(10):3628–3632
24. Brack C et al (1978) A complete immunoglobulin gene is created by somatic recombination. Cell 15(1):1–14
25. Janeway CA Jr (1989) Approaching the asymptote? Evolution and revolution in immunology. Cold Spring Harb Symp Quant Biol 54(Pt 1):1–13

Further Reading

Knoll AH (2011) The multiple origins of complex multicellularity. Annu Rev Earth Planet Sci
    39:217–239
Lynch M (2007) The frailty of adaptive hypotheses for the origins of organismal complexity. Proc
    Natl Acad Sci U S A 104:8597–8604
Sebe-Pedros A, Degnan BM, Ruiz-Trillo I (2017) The origin of Metazoa: a unicellular perspective.
    Nat Rev Genet 18(8):498–512
Wolpert L (1992) Gastrulation and the evolution of development. Development 161(Suppl):7–13

Chapter 3
Innate Immunity

3.1 Structure of the Innate Immune System

All defence systems have much in common. No matter whether we are thinking in terms of defending a mediaeval castle from greedy neighbours or a multicellular eukaryote from attack by pathogens, the basic story is always the same. The defence system employed will consist of three parts. The first part provides information about whether a dangerous situation is developing. In the case of the castle, sentries will provide this necessary information. In the case of an animal's innate immune system, soluble extracellular receptor molecules, and cell-associated sensors expressed by macrophages and other innate sentinel cells will detect incipient infections or other deviations from the homeostatic norm.

The second part of the defence system is a command and control module that analyses the information provided by the sensors, and then reaches a conclusion as to what—if anything—now has to be done. In the case of the castle, this function is carried out by the commanding officers. In the multicellular eukaryote it is the signal transduction chain connecting the sensors on the sentinel cell's surface to its output, together with the interactions between the various cells involved in the response that provides the information processing part of the system.

The third and final component consists of the weapons, which will be used to destroy the enemy. In the castle, these might be cannon, or troops of mercenaries. In the innate immune system the hardware consists of things like hydrolytic enzymes or reactive oxygen intermediates, and these have changed little from what was available to an amoeba hundreds of millions of years ago. These terminal effectors are, in principle, capable of dealing with any and all pathogens, but since immunity is a dispersed system, mechanisms must be available to concentrate them at the site of an infection, and great care must be taken in their application, for if inappropriately activated they will destroy host tissues.

This basic structure implies that both in the castle and in the multicellular eukaryote the defence system is modular. In the castle, the sentry is a separate entity

© Springer Nature Switzerland AG 2019
R. Jack, L. Du Pasquier, *Evolutionary Concepts in Immunology*,
https://doi.org/10.1007/978-3-030-18667-8_3

from the cannoneer, and in innate immunity the sensor molecules and terminal effectors are, by and large, separate molecular categories. This is indeed the general rule—but in biology there is always at least one exception. In this case the inevitable exception appeared very early in eukaryotic evolution for it is already present in the single cell amoebae [1]. It takes the form of a "smart" weapon, which once released can autonomously set out to search for and then destroy intruders. For this to work, the information necessary to identify the pathogen, as well as that needed to destroy it, has to be built into the structure of the molecule, and that information has to be unambiguous. This is the way that constitutively expressed cationic antimicrobial peptides (CAMPs) work. They will associate with the anionic microbial surface, and then the CAMP's hydrophobic domain invades the microbe's surface membrane. Such constitutively produced CAMPs are extensively used in the innate immune systems of invertebrates. In vertebrates their role in defence is less prominent, since they are confined largely to mucosal surfaces. Apart from CAMPs, the sentinel functions and the effector functions of innate immunity are generally in separate molecules, and these are linked only once the sentinel has engaged its target.

3.2   Evolution of Innate Immune Receptors

As described in Chap. 2 immunity consists of two arms—innate and adaptive—and there are many differences between the ways that the generation of receptors is organised in these two arms. This chapter will be concerned with the receptors of innate immunity, while those of adaptive immunity will be discussed in Chap. 4. However, it is worth noting here that one of the most striking differences between these two immune defence strategies is that the ligand-binding domains of the receptors of innate immunity are formed from a plethora of different protein structures—Leucine Rich Repeats (LRR), C-type lectin domains, helicase domains, pentraxin domains and so on (see Appendix B)—while in the protein-based adaptive immune systems in vertebrates, all of the antigen-specific receptors are members of a single protein family. In the non-jawed vertebrates the adaptive system receptors are formed from concatenations of LRRs; in the jawed vertebrates all of the antigen-specific receptors are formed from the so-called Immunoglobulin Superfamily (IgSF) domains. Why is such a broad array of receptor structures exploited in innate immunity, while each of these adaptive systems relies on just one single structural form? This difference is all the more striking given that the same fundamental processes—random mutation and selection—formed both the innate and the adaptive immune system receptor repertoires. The answer may simply be that the trade-offs required for the evolution of innate and adaptive immune systems are not identical. Imagine the evolution of an innate defence system: no single receptor will be able to detect all possible pathogens, and the number of different receptors is limited by the lack of space in the genome. Because of this, the choice of each member of the innate receptor repertoire becomes a matter of life or death. How is the collection of receptors to be assembled? If random mutation tinkers with some gene in such a way that its product now acts as a receptor that provides protection against some type of pathogen, then this successful mutant gene will be positively selected. Mutations are random and are happening all the time, and millions of years were available to slowly accumulate the necessary changes. Over this long time span many rounds of mutation and selection could take place. Many different genes could be tinkered with, and many different potential receptors made available for selection. This is just as well, for having a receptor repertoire composed of many different molecular forms makes it so much harder for a pathogen to come up with a virulence strategy that will counter them all, and hence disarm the entire repertoire in one fell swoop.

The evolutionary trajectory followed by adaptive systems is very different. Here billions of different mutant versions of just one single receptor form provide a repertoire that is able to detect every conceivable pathogen. However, these mutants do not slowly accumulate over millions of years. Instead, their formation has to take place in just a few short hours during lymphocyte differentiation. Were the millions of years' worth of evolutionary change to be squeezed into these few hours, then the mutational load on the lymphocyte would be lethal. In order to survive, the lymphocyte has to use a special somatic mutational process that targets the changes to the particular, defined genetic locus encoding the receptor complex. Mutations in other loci, and chromosomal rearrangements of all sorts, have to be kept to the absolute minimum, for otherwise the cells would die or be transformed. A special mutational mechanism therefore had to be established—one that can be restricted to a particular locus, and this locus has to be structured so that it becomes a suitable target of that mutational process. This drastically limits the number of different target loci and hence the number of different types of receptor that can be expressed.

There is one further crucial difference between innate and adaptive systems. The innate immune receptors have been selected for their ability to increase fitness by combating pathogens. They form a germline encoded "phylogenetic memory" of the pathogens that have been encountered during the course of evolution. These receptor genes, fixed in the genome, are immediately available to all of the cells involved in an immune response. The response is consequently very fast. This is fundamentally different to the way things are done in adaptive systems, where somatically formed receptors are clonally restricted to a tiny number of cells that have to respond to activation with a period of intense proliferation. Primary adaptive immune responses are consequently slow in comparison to innate responses. These matters will be discussed further in Chap. 4.

Some innate immune system receptors are listed in the table in Appendix B. One can see here two general themes in their evolution. First, the gene for a successful receptor structure is often repeatedly duplicated, and the new copies then randomly mutated to generate receptors with new binding specificities. In this way families of receptors are formed in which the different members recognise different ligands. Second, many receptors consist of two or more fused functional modules. Before turning to a few examples of these receptors we will briefly outline the genetic processes—gene duplication and exon shuffling—that lie behind their generation.

3.2.1   Gene Duplication

If a gene is duplicated, then one copy may be positively selected to retain the original function, while random mutation and selection may give the other a new and useful function. If this happens, then the new gene may become fixed in the genome. Since this is so, why doesn't the number of genes not just keep on getting larger and larger? The answer is that most duplicated genes fail to find a new function, and instead rather quickly degrade; first to recognisable "pseudogenes" and then, as ever more mutations accumulate, they gradually lose their identity. However, in rapidly evolving systems such as immune defence, genes—indeed whole families of genes—are born, expand, contract and die at quite astonishing rates. What mechanisms permit gene families to undergo such "yoyo"-style expansion and contraction cycles?

Perhaps the most dramatic mechanism for gene expansion involves a whole genome duplication (WGD) in which every gene in the organism is doubled. This happens relatively frequently in plants, but in most animals it is rare, and when it does happen it is usually not very successful. Only rather few ancient WGDs have persisted. Nevertheless, though they seem not often to survive in animals, those that do then give rise to evolutionarily successful lineages. Two ancient WGDs occurred in the vertebrate lineage between 550 and 600 million years ago and a third, much later WGD event, took place in the line leading to bony fish. From that point on, WGD in vertebrates has been restricted to fish, amphibians and reptiles.

Successful WGD in other vertebrates are so infrequent that they cannot explain the expansions and contractions of gene families that take place rapidly and repeatedly throughout the evolution of vertebrates. These frequent gene copy number changes rely on a more prosaic mechanism that involves the vast numbers of repetitive sequence elements that make up over 50% of the human genome [2]. They provide the opportunity for misalignments during chromosome pairing and recombination in meiosis I, and so for the expansion or contraction of gene copy number (Fig. 3.1).

3.3   Evolution of a Gene Family by Duplication and Modification: the Toll-Like Receptors

We can take the Toll-Like Receptors (TLRs) as an example of a gene family that has expanded and contracted over the course of evolution, acquiring new characteristics along the way. In the nematode worm *Caenorhabditis elegans* there is just one member of the family present, in the purple sea urchin *Strongylocentrotus purpuratus* there are over 200, while in mammals there are usually around 10 [3]. What makes the TLR family of particular interest is not just the fact that the number of family members rises and falls quite dramatically across phylogeny, but also that this family serves to illustrate how random mutation and selection, acting on the products of a gene duplication event, can give rise to a family of

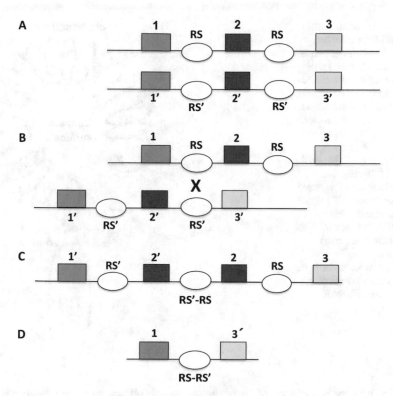

Fig. 3.1 Gene copy number changes in meiosis 1. (**a**) Section of a chromosome carrying the genes 1, 2 and 3 normally aligned with its homolog carrying the alleles 1′, 2′ and 3′. Copies of a repetitive element (RS) are present flanking gene 2 and 2′. (**b**) Pairing misalignment of the repetitive sequence elements in meiosis 1 with a crossover (X) in the repetitive elements. (**c**) Recombination product with a duplication of gene 2 (expansion). (**d**) Recombination product with a deletion of gene 2 (contraction)

receptors whose members have remarkably divergent-binding characteristics. The mammalian TLRs all have a ligand-binding domain, which is composed of a head-to-tail concatenation of a small protein fold known as the LRR. LRRs are short protein domains, typically 20 to 30 amino acids long, which, when joined together head to tail, produce a horseshoe-shaped protein structure (Fig. 3.2). Many thousands of proteins containing such concatenations of LRR domains are known, and they are found all the way from bacteria to mammals. The function of most is unknown, though it is thought that in many cases the LRR domain is involved in protein–ligand interactions. The first crystal structure of an LRR protein–ligand complex was that of the pancreatic ribonuclease inhibitor bound to its ligand—ribonuclease-5. In this complex, the interacting surface of the ribonuclease-5 fits inside the horseshoe structure formed by the LRR domains of the ribonuclease inhibitor, and as a result, the surface area involved in the interaction is large—double the area involved in a typical antibody–antigen interaction [4]. This mode of binding

Fig. 3.2 Schematic outline of a Toll-Like receptor (TLR). The extracellular part of the TLR molecule shown consists of Leucine Rich Repeat (LRR) domains (open ovals), which form a horseshoe-shaped structure that is "capped" by an N-terminal domain (NT-cap) and a C-terminal domain (CT cap). The intracellular domain couples the TLR to the cell's signal transduction machinery. This "TIR" (TLR and interleukin-1 receptor) domain is found in homologous form also in the intracellular domain of the interleukin-1 receptor

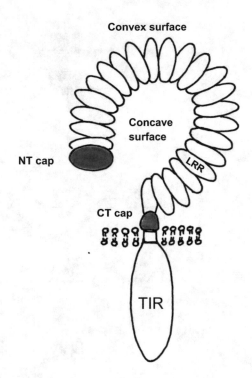

explains why the association of the ribonuclease inhibitor with ribonuclease-5 is one of the tightest non-covalent interactions known in biochemistry. Many other LRR proteins share a similar overall structure, and bind their ligands in a similar way. These include the antigen-binding molecules produced by the VLRA, VLRB and VLRC cells of the lamprey adaptive immune system (see Sect. 4.3) and the eponymous *Drosophila* protein "Toll-1".

3.3.1 Drosophila *Toll-1*

Toll-1 is a *Drosophila* cell surface transmembrane protein whose extracellular ligand-binding domain is formed of LRRs that gives it a horseshoe-shaped structure. During embryogenesis of the fly, Toll-1 contributes to the definition of the position of the dorsal-ventral axis by binding an endogenous protein called "spätzle". When spätzle is bound, Toll-1's intracellular signalling domain is activated, and this leads to the activation of the transcription factor NFkB, which in turn leads to the changes in the gene expression pattern that are necessary to impose the dorsal-ventral axis. The involvement of Toll-1 and spätzle in this all-important decision is, however, a hit and run affair, for once the dorsal-ventral axis has been set, Toll-1 is no longer required to enforce it. The Toll-1 gene is thus free to find some other useful employment, and this it has done by "moonlighting" as an essential part of the signalling cascade, which provides the fly with protection against fungal and Gram-positive bacterial infections. In this second setting, the detection of pathogens by *Drosophila* sensor molecules sets off a protease cascade, at the end of which pro-spätzle in the haemolymph is converted to the mature spätzle form, which then binds to Toll-1 expressed on the fly's immune sentinel cells. This interaction generates a signal that activates the transcription factor NFkB, and results in the production of antimicrobial peptides. The crystal structure of the complex of Toll-1 bound to spätzle has been solved, and once again—as in the case of the pancreatic ribonuclease inhibitor—the ligand is bound to the inner face of the horseshoe [5]. From all of this, a picture emerges of proteins with LRR domains: they form horseshoe-shaped structures, and their ligands bind on the inner (concave) face, where there is a large surface area available for making productive molecular interactions.

3.3.2 Mammalian Toll-Like Receptors

There is, however, a group of proteins containing concatenated LRR domains, which have been subject to selective pressure that has completely changed their mode of binding ligands. These exceptions are the TLRs of vertebrates, which play important roles as innate immune system sensors. There are ten TLR genes in humans all of which code for proteins that consist of an extracellular concatenation of LRR domains and an intracellular signalling domain that couples ligand binding to the activation of the transcription factor NFkB. They get their name from the structural similarity of both the extracellular LRR and the intracellular signalling domain to those of Toll-1 in *Drosophila* (Fig. 3.2). However, in the case of the mammalian TLRs the rules governing LRR protein–ligand interactions are turned on their head, for TLRs do not bind their ligands on the inner (concave) face, but rather on the outer (convex) face, or on the side of the horseshoe. Different TLRs bind their ligands using quite different parts of the horseshoe surface.

What might be the selective advantage of abandoning the tried and tested way of binding ligands to the concave face of the horseshoe, and instead binding them on the convex face? We really don't know, but two types of answer have been offered. The first has to do with the structure of the LRR. Each LRR repeat is composed of around 25 amino acids of which the first 10 to 11 are quite highly conserved, and they form the structurally important inner face of the horseshoe. The outer face of the horseshoe is, in contrast, built of the less conserved second halves of the LRR repeat sequences. The convex face therefore accommodates much more sequence variability and thus permits the selection of a wide range of ligand-binding options. This may explain the evolution of the different members of the TLR family, which bind such very different types of ligand molecules. For example, in the mouse TLR-1, -2 and -6 are involved in binding bacterial lipopeptides, TLR-3 binds double-stranded RNA, TLR-4 is required to bind the lipopolysaccharide of Gram-negative bacteria, while TLR-5 binds bacterial flagellin. TLR-7 and -8 bind single-stranded RNA, while TLR-9 binds double-stranded DNA.

The use of the outer face of the horseshoe for ligand binding by the TLRs also has a further advantage. Binding of a large protein ligand to the inner—concave—face blocks access of any further ligands to the "inside" of the horseshoe. In contrast, binding of a ligand on part of the outer face leaves open the possibility of a simultaneous interaction with further ligands, and this sort of thing has now been demonstrated both for TLR7 and for TLR8 [6]. A TLR molecule, which must interact with two different ligands for activation, plays the role of an "and" device in the logic circuit governing activation of innate immunity.

3.4   Exons and Introns: Making New Proteins by Swapping Domains

Duplications are not the only route to new genes and proteins. The curious arrangement of the coding regions of many proteins in eukaryotes provides another means to rapidly generate novel protein structures by mixing functional domains "borrowed" from already existing genes. In prokaryotes, genes are encoded in one continuous stretch of DNA sequence, but in eukaryotes the coding regions of almost all genes are interrupted with non-coding sequences referred to as "introns". These introns are typically very much longer than the coding "exons". The entire stretch of DNA encoding the gene—introns as well as exons—is transcribed into one continuous piece of RNA. The introns then have to be precisely removed from the transcript to form the mature messenger RNA, which can be translated into protein. Why should such a complicated system have been of selective value in the evolution of eukaryotes? There are two major advantages offered by doing things this way. The first is that it allows the effective rate of evolution to be speeded up enormously. The second is that it provides a means of extending the available genetic information.

3.4.1   Exon Shuffling: Speeding Up the Rate of Acquisition of New Genes

The way in which genes are divided into exons and introns is not random. Particularly in vertebrates, the exon stretches are frequently correlated with the borders of functional domains in proteins [7]. Because of this, new genes can be constructed simply by recombining pre-existing exons "borrowed" from other genes in the genome. This shuffling of exons allows the formation of new genes at a rate that is orders of magnitude faster than could be achieved by mutation and selection acting on unselected DNA sequences. In the example schematically outlined in Fig. 3.3, an

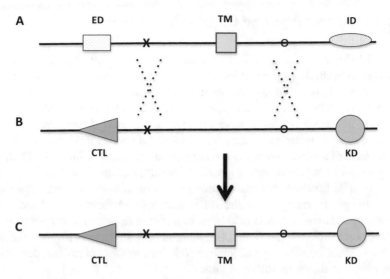

Fig. 3.3 Exon shuffling generates new proteins with new functional properties. (**a**) A gene coding for a membrane protein. The gene consists of three exons. The first codes for an extracellular domain (ED) (open rectangle), the second for a transmembrane domain (TM) (shaded square) and the third for an intracellular domain (ID) (stippled oval). (**b**) An unrelated gene with two exons codes for an intracellular signalling protein. The first exon codes for a ligand-binding C-type lectin domain (CTL) (grey triangle). The second exon codes for a kinase domain (KD) (grey circle), which can phosphorylate certain signal transduction protein targets. The two genes are unrelated and share only repetitive sequences X and O in their introns. In rare cases these may permit recombination (dotted lines). (**c**) The result of such a recombination event is a gene in which a CTL encoding exon is linked to one coding for a transmembrane domain and then to the kinase domain

intracellular protein involved in signal transduction is converted in one step into a membrane protein, which may now function as a receptor.

Traces of this process can be seen in many innate immune receptors (see Appendix B). We can take, for example, the Mannose-Binding Lectin (MBL), which is made up of a mannose-binding C-type lectin domain coupled to a collagen domain (Appendix B). C-type lectin domains are found in a large number of different proteins all of which are members of the "C-type lectin" family. The important point here is that not all proteins containing a C-type lectin domain also contain a collagen domain, and vice versa, not all proteins with a collagen domain have a C-type lectin domain. The MBL arose when an exon coding for a C-type lectin domain was "shuffled" in the genome to within striking distance of an exon encoding a collagen domain. This illustrates that once a functional domain of a protein has evolved, it does more than just code for part of one particular protein; it also serves as an evolutionary asset that can be shuffled around in the future to quickly produce new proteins with diverse functions.

3.4.2 Exon Splicing: One Gene—More Products

The second major advantage of having coding sequences interrupted by introns is that after transcription of a gene, the splicing of the RNA transcript may be organised so that not only introns but also one or more exons are removed. The result is that one gene may yield more than one mRNA sequence. This process of "alternative splicing" is not just a matter of random error, but is used in particular cells in a directed fashion to favour one or the other product. Of course, if the various different splice variants of a gene were all to do more or less the same thing, then alternative splicing would be no more than a genetic curiosity, but this is not the case. Most proteins in cells function as parts of protein networks, and different networks do different things. In many cases the differently spliced forms of a human gene physically interact with partners belonging to different networks, and hence presumably have quite different functions [8]. Alternative splicing thus extends the available genetic space by allowing one gene to code for a number of distinct polypeptide chains, which do distinctly different jobs.

3.4.3 Adaptive Introgression

Exon shuffling involves borrowing pre-formed parts of one's own genes to make new functions. Introgression involves borrowing pre-formed genes from some other species. Species are notoriously hard to define (see Sect. 1.2), and one reason for this is that the reproductive boundary between two species may sometimes be somewhat "leaky". Introgression happens when two species, "aa" and "bb", mate to produce an "ab" F1 hybrid. If the F1 is fertile and is serially backcrossed to the "aa" parental species, then the "b" genetic material will be gradually diluted out. However, if a "b" gene offers some adaptive advantage, then it may be positively selected and become incorporated into the "aa" background. Introgression is relatively common in plants, but much less so in animals. It has, however, been studied in a number of groups ranging from amphibians to mice—with humans having become the most recent species of interest.

It is now widely accepted that humans, having left Africa, met and mated with archaic hominid species—Neanderthals and Denisovians—in Eurasia, and that this resulted in archaic sequences being introgressed into human genomes [9]. One example of such introgression concerns the region containing the gene coding for the transcription factor EPAS1. In Tibet—but only in Tibet—some 80% of the population carries an EPAS1 haplotype derived by introgression of a Denisovian segment of DNA. Because this archaic haplotype is correlated with increased haemoglobin levels that are advantageous for life at high altitudes, this is considered to be a case of adaptive introgression. A second example centres on a 143 kilobase pair stretch of DNA containing the genes coding for the innate immune receptors TLR-6, TLR-1 and TLR-10. There are seven known haplotypes of this "TLR6, 1, 10" segment in the genomes of present day humans. Three of these have an archaic origin—two being derived from Neanderthals and one from Denisovians. The Neanderthal-derived haplotype III is present at frequencies of 11% to 51% in non-African populations, which suggests that it may well have been positively selected. The nucleotide exchanges between the major "human" haplotype and the "Neanderthal" haplotype III do not involve changes to the coding regions of the TLR genes. Instead the differences are restricted to the non-coding regions, but these exchanges overlap with transcription factor-binding sites and are correlated with increased expression of these TLR genes. The Neanderthal-derived haplotype is associated with altered immune phenotypes, including a reduced seroprevalence of *Helicobacter pylori* and with increased susceptibility to allergies. This Neanderthal haplotype may have been advantageous to different extents in different populations facing successive waves of pathogens, and thus its introgression may also be the result of positive selection [10].

3.5 Extracellular Innate Immune Receptors and Their Targets

In an ideal world the innate immune receptors should detect molecular structures that are only associated with pathogens (Pathogen Associated Molecular Patterns—PAMPs), but they should never interact with "self" structures. Our world, however, is far from ideal, and so the repertoire of innate system receptors is not, and cannot be, perfect in this sense. Nevertheless, the collection of innate system receptors is indeed remarkably effective against pathogens, and these receptors have been naturally selected as a "phylogenetic memory" over many millions of years to be "safe" with respect to our own healthy tissues.

A long list of the innate system's diverse collection of receptors, and an even longer list of their ligands would probably not be terribly informative, and would certainly make for a hard read. However, to get an idea of what is involved, one can perhaps briefly take some examples of the sorts of structures that innate immunity targets, and the types of receptors that have evolved to detect them (see Appendix B). In this respect, perhaps the most important point to bear in mind is that the number of pathogens that may evolve is unlimited, and yet the number of genes available to counter them, and hence the number of different germline-encoded receptors which can be maintained, is very strictly limited. Because of this, innate immunity concentrates on structures that are shared by a large number of different microorganisms. The structures targeted should be essential for the survival or for the virulence of the pathogen, so that the microorganism cannot evade detection simply by discarding them, and if, in addition, eukaryotic cells cannot synthesise the targeted structures, so much the better, for then there will be small chance of an innate attack being initiated against "self". Yet the term "self", in this context, has to include not just the host's eukaryotic cells, but also all of the commensal microbes that inhabit mucosal and other surfaces. Targeting ligands that are present on a broad range of microorganisms suffers from the drawback that these structures are not restricted to pathogens, and of course turning the arsenal of immunity's terminal effectors indiscriminately on, for example, intestinal commensals would lead to disaster. Ways have had to be found to discriminate host from microbe, and to distinguish dangerous pathogens from harmless commensals [11]. A few examples will serve to illustrate the ways in which evolution has faced these challenges.

3.5.1   Receptors that Target Surface Structures on Extracellular Microbes

Perhaps the best-studied examples of receptors that target structures expressed only on microbes are the innate immune receptors that detect prokaryotic-specific features of the bacterial outer surface. In Gram-negative bacteria the cytosol is bounded by a membrane, outside of which lies a thin cell wall consisting of cross-linked peptido-glycan. Beyond this cell wall, there is a so-called outer membrane, and this is essential for the bacterium's survival. This outer membrane has an unusual structure in that the major component of its outer leaflet is a glycolipid called lipopolysac-charide (LPS). LPS would seem to be an almost perfect target for innate immunity since it is not produced by eukaryotic cells, and so there is no way that a response against it can be directed at our own healthy tissues. Furthermore, since all Gram-negative bacteria have such an LPS-containing outer membrane, a receptor able to detect it provides innate immunity with a means of sensing the presence of any of a huge range of different bacterial species, Finally, Gram-negative bacteria make no secret of their presence, for they constantly release LPS during growth and division. However, in host–pathogen interactions things are seldom simple, and there is a problem with LPS—one that nicely illustrates evolution's ability to use random mutation and selection to overcome seemingly intractable difficulties. The problem is that the LPS molecule is composed of two quite different elements. The first of these is a long hydrophilic polysaccharide chain, and these chains cover much of the surface of the bacterium. The structure of these chains varies considerably between different bacterial strains, so much so, that antibodies raised against them can be used to type Gram-negative bacterial species. Because of its great variability, this hydrophilic part of the LPS molecule is not a suitable target for innate immunity. However, the variable polysaccharide is covalently attached to a hydrophobic lipid core called lipid A (Fig. 3.4), and this has a highly conserved structure. The problem, however, is that the physical properties of the LPS molecule—hydrophobic at one end, hydrophilic at the other—means that LPS released into an aqueous environment during bacterial growth immediately forms micelles in which the hydrophobic lipid A part is all tied up in the inside, while the hydrophilic polysaccharide chains face the aqueous phase. The LPS is there for all to see, but the all-important lipid A part, being hidden inside the micelle, is completely invisible to a receptor in the aqueous

Fig. 3.4 The lipopolysaccharide of Gram-negative bacteria. (**a**) The LPS molecule is composed of a hydrophobic Lipid-A (LA) core attached to a hydrophilic polysaccharide (PS) chain. (**b**) The Gram-negative bacterial outer membrane consists of two leaflets. The outer leaflet is composed of phospholipids (PL) and LPS while the inner leaflet is composed of phospholipids only. (**c**) LPS released from the membrane forms micelles in aqueous solution. In these micelles the hydrophobic Lipid-A part is hidden within the micelle

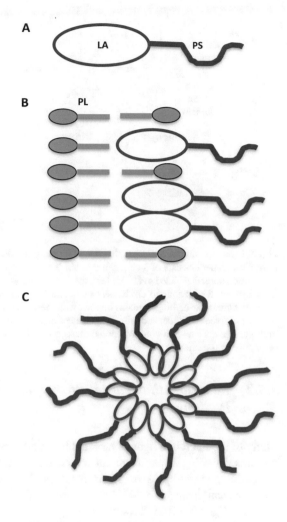

phase (Fig. 3.4). Evolution's solution to this problem has been complicated. It provides us with a protein called Lipopolysaccharide-Binding Protein, which extracts LPS from the micelles and passes it on to the intermediate carrier CD14, which in turn passes it on to MD-2, a protein that is non-covalently associated with the cell surface molecule TLR-4 expressed on innate sentinel cells. TLR-4 then signals the presence of Gram-negative bacteria into the cell. This represents a considerable genetic investment, for four different proteins are involved in this binding cascade, but this investment paid off in terms of fitness, because it provides us with the capacity to detect a very wide range of different bacterial pathogens.

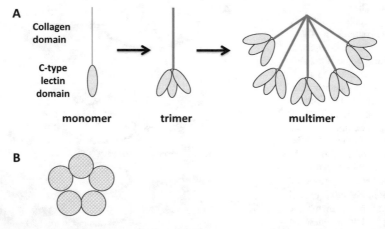

Fig. 3.5 Innate immune receptors that detect ligand patterns: (**a**) Mannose-binding lectin (MBL). Left: a monomer consists of a collagen domain (solid line) linked to a C-type lectin carbohydrate recognition domain (shaded oval). Middle: the monomers associate through the collagen domains to form trimers. Right: four to six trimers associate to form the receptor. Activation requires that several of the sugar-binding domains interact with their ligand. (**b**) The Pentraxin C-reactive protein consists of five saucer-like subunits. Each saucer has a ligand-binding site on one face. Binding to multiple ligands causes a conformational change that forms a complement activating site on the other face of the subunit

3.5.2   Receptors that Target Prokaryotic Surface Molecular Patterns

Many of the microbial ligands targeted by innate immunity are not really "safe", in the sense of being absent from eukaryotic cells, for many innate immune receptors recognise phospholipids or sugar residues that are present on all cells. These receptors can distinguish "self" from "non-self" because the patterns of expression of these ligands differ between eukaryotic cells and microbial surfaces. MBL (Fig. 3.5) is an example of this sort of thing. This member of the "collectin" family of soluble receptors is composed of 12 to 18 identical subunits each of which is terminated in a sugar recognition domain that binds mannose residues, albeit with rather low affinity. Large numbers of mannose residues are accessible on the surface polysaccharides of microorganisms, while the surface glycoproteins of mammalian cells carry polysaccharide chains that are usually terminated with galactose or sialic acid—sugars to which MBL does not bind. Nevertheless, the odd mannose residue will certainly be accessible on the surface of mammalian cells, and MBL must be able to ignore them. This is where the question of ligand distribution becomes important, for MBL will only be activated if many of its binding sites interact with mannose residues. The structure of the MBL molecule thus defines the area on a surface within which these mannose residues must appear. This sort of distribution of mannose residues is typical of the surface of many bacteria, but not of the surface of mammalian cells.

The pentraxin C-reactive protein (CRP) (Fig. 3.5b) is another soluble receptor that operates in a rather similar way. It is a pentamer, and each of the identical saucer-shaped subunits recognises a phospholipid structure. Once again multiple ligand interactions are required for activation and the required ligand distribution is defined by the spatial separation of the five saucers. Activation of MBL or of CRP leads to activation of the complement system (see Sect. 3.9.2), one of the key effector arms of innate immunity.

3.5.3   Provision of Endogenous Bait

So far we have looked at some examples of the sorts of receptor categories that directly detect molecular components of pathogen structures. This is the central means employed by innate immunity to counter pathogen attack. There are, however, alternative ways of doing things. One of these is the provision of endogenous bait molecules. The object of the exercise here is to recognise that a particular type of virulence mechanism is being applied—irrespective of which particular pathogen is involved. To do this the host offers as bait a molecule that will be destroyed by a certain type of virulence mechanism, and destruction of the bait then triggers a host response. This is an immune strategy that is widely used in plants, though it is less central in animal immune defence. One example of this sort of thing has been demonstrated in the chicken intestine. The intestinal lumen is packed full of bacteria, viruses and fungi most of which are completely harmless, and since these are held at bay by the layer of mucous that covers the intestinal epithelium they may, by and large, be safely ignored. However, microorganisms that penetrate the mucous layer and approach the epithelium must be considered to be a potential threat. Many pathogenic bacteria produce proteases that help them to breach the epithelial barrier, and in the chicken the TLR-15 receptor has been selected to detect them. TLR-15 is expressed on the luminal side of the chicken intestinal epithelial cells but, unlike other TLRs, it does not bind a microbial structure. Instead it is activated when its extracellular domain is destroyed by proteolysis. It is protected by the intestinal mucous layer from proteases released by the harmless commensal flora in the lumen [12], but pathogenic yeast or bacteria that penetrate the mucous layer, swallow the TLR-15 bait, and by doing so induce a local inflammatory response [13].

3.6   Intracellular Innate Immune Receptors and Their Targets

While bacteria and fungi are major targets of those innate receptors that are directed to the extracellular space, by far the most important intracellular pathogens are viruses. They pose a special problem for the innate system because they lack widely distributed common surface targets equivalent to the LPS of Gram-negative bacteria. This is not to say that receptors detecting surface features of viruses are never used; however, it is not the major innate system approach to counter viral attack. Instead, innate immunity has learned to target the only thing that all viruses share in common—a nucleic acid genome. This in turn has required the evolution of intracellular receptors, because the viral genome is only accessible once the virus has infected a cell. The problem with using nucleic acids as identifiers of pathogens is that most host cells also have both a nuclear and a mitochondrial genome composed of DNA, and they synthesise large amounts of RNA. Receptors detecting nucleic acids must be able to clearly distinguish host nucleic acids from those derived from the pathogen. That is not so easy, for the plain fact of the matter is that, in terms of general structure, one piece of DNA or RNA looks much like any other. In this situation the detection systems have to have evolved the capacity to make use of extra information in order to make the all-important distinction between "self" and "non-self". This extra information is not dependent on the sequence of bases in the nucleic acid, but it may involve questions of cytoplasmic geography, ligand concentration or nucleic acid fine structure.

3.6.1   Sensing the Location of Nucleic Acids: APOBEC3G

In eukaryotic cells, DNA is normally expected to be present in large amounts in only two locations—in the nucleus and in the mitochondria. If large amounts of DNA suddenly appear free in the cytosol, then that is a clear indication that something has gone badly wrong. Retroviral infection is a case where single-stranded DNA is generated in the cytosol, and the APOBEC3G deaminase has evolved in mammals as an innate system effector that deals with this particular situation. Retroviruses are single-stranded RNA viruses. On entering the cell the RNA is translated to produce the viral reverse transcriptase, which makes a DNA copy (cDNA) of the RNA genome. The viral genome is now a RNA:cDNA hybrid. The RNA strand of the hybrid is then destroyed so that the single-stranded cDNA can be converted into double-stranded DNA (dsDNA), and so for a brief moment the viral genome exists in the cytosol as a single-stranded cDNA molecule. This moment is seized on by innate immunity, which uses a member of the APOBEC3 family of deaminases, to convert deoxy cytidine (dC) residues in the single-stranded cDNA into deoxy uridine (dU). By inserting lots of dC→dU mutations into the viral genome many of the virus particles that will be produced are no longer biologically functional [14].

   This way of introducing mutations into DNA by using a cytidine deaminase is worth bearing in mind, for a much more ancient member of this APOBEC family, "Activation-Induced Deaminase" (AID), arose at, or close to, the point where the first vertebrates evolved [15], and AID plays central roles in the vertebrate adaptive immune systems to be discussed in Chap. 4.

3.6.2   Sensing the Local Concentration of DNA: cGAS and STING

APOBEC3G is far from being the only mechanism for detecting pathogen-derived nucleic acid in the cytosol of eukaryotic cells, and one of the most important innate mechanisms to combat viruses infecting the cytosol of vertebrate cells centres on the synthesis of a cyclic dinucleotide (see Appendix C). The protein initially involved in this process—"cyclic GMP-AMP synthase" (cGAS)—is an enzyme that binds DNA, and having done so then synthesises the cyclic dinucleotide cyclic GMP-AMP (cGAMP). The cGAMP that is produced is then bound by a second protein called "Stimulator of Interferon Genes" (STING). Once STING has bound the cGAMP it activates the expression of interferon genes. These interferon genes arose early in vertebrate evolution, and they code for secreted molecules that function as cytokines. They bind to ubiquitously expressed interferon receptors on cells, and by doing so induce a large number of genes that are powerful mediators of antiviral responses. The interferon system is thus central to innate antiviral defence in vertebrates [16].

The evolution of the vertebrate cGAS—STING system nicely illustrates the theme of "evolutionary borrowing". First, vertebrates inherited the genes for both cGAS and STING from their invertebrate predecessors. Second, early in vertebrate evolution cGAS "borrowed" a zinc ribbon domain by exon shuffling, and it is this domain that allows it to bind DNA, and so converts the vertebrate enzyme into a cytosolic DNA sensor. In contrast, invertebrate cGAS enzymes do not have this domain, cannot be activated by binding DNA and hence cannot act as DNA sensors. The means of inducing these invertebrate cGAS enzymes is currently unknown (Fig. 3.6). The third level of borrowing concerns STING—and it is perhaps the oddest of them all. As noted above STING is an activator of interferon genes and, since interferon genes are a vertebrate "invention", any reference to an invertebrate STING is something of a misnomer. Indeed, though a STING homolog is present in the sea anemone it is quite unclear what it does once it has been activated by binding a cyclic dinucleotide. However, in *Drosophila* the situation is somewhat better understood. Though we know few details about it, it is nevertheless clear that *Drosophila* STING plays a role in defence against pathogens. Once activated it can induce the expression of antimicrobial peptides [17], and it is also involved in defence against certain viruses [18]. Thus, the immune involvement of the invertebrate cGAS-STING module has been retained in vertebrates, but tinkering has led to crucial changes in both the induction and output components: cGAS has been altered to make it responsive to DNA, while STING has been altered to make it an inducer of the vertebrate-specific interferon genes.

The ability of vertebrate cGAS to act as a pathogen sensor depends on there being no detectable DNA in the cytosol of healthy cells. Sadly, however, an assertion that host DNA is never found in the cytosol of healthy cells would be misleading. Small amounts of DNA fragments, derived from nuclear or mitochondrial replication intermediates, do indeed find their way into the cytoplasm. Their concentration is

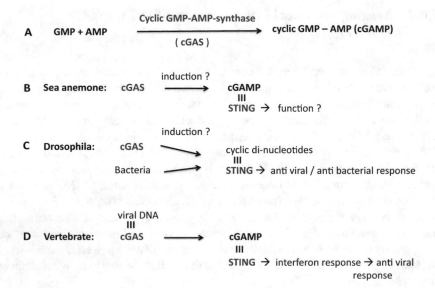

Fig. 3.6 cGAS and STING. (**a**): The enzyme cyclic GMP-AMP synthase (cGAS) forms a cyclic dinucleotide (cGAMP) consisting of one GMP moiety linked to an AMP moiety. (**b**) In the starlet sea anemone cGAS may produce cGAMP, which activates STING. Neither the means of inducing cGAS nor the function of the activated STING is known. (**c**) In *Drosophila* STING is involved both in antibacterial and in antiviral responses. Cyclic nucleotides derived from bacterial pathogens may directly activate STING. Whether *Drosophila* cGAS is involved in these responses is currently not known. (**d**) In vertebrates cGAS can bind to viral DNA and respond by producing cGAMP to activate STING, which in turn activates the production of type 1 interferons that initiate a broadly effective antiviral response

low, but it is not zero. Because of this, the cGAS in the cytosol has to be able to ignore this normal level of "self" DNA, and that is not so easy, for cGAS has no capacity to distinguish host from viral DNA on the basis of sequence—it simply binds to the sugar phosphate backbone of any DNA that it may meet. What is needed is a means of setting a threshold so that cGAS ignores the low levels of endogenous DNA in the cytosol, but responds to the increased levels present during a viral infection. The threshold seems to be set by a cytoplasmic nuclease called Trex-1, which is able to destroy the "normal" levels of endogenous DNA, but is overwhelmed by the products of viral infection [19]. In this case the innate immune receptor fails to distinguish structurally between the ligands from host and pathogen, but rather reacts to a change in concentration of a ligand that is common to both.

3.6.3   Sensing RNA Fine Structure: RIG-I

An even more awesome challenge faces innate immunity's receptors that must detect viral RNA in the cytosol. As with the receptors for cytosolic DNA, they cannot distinguish "self" sequences from pathogen ones, and hence they are not inherently pathogen-specific. Instead they must recognise some general, but sequence-independent, pathogen-specific aspects of nucleic acid structure. This is a particularly acute challenge, for the cytosol is full of host RNA of one sort or another. Ribosomal RNA, transfer RNAs, micro RNAs and messenger RNAs are all host-derived, and all are present in the cytosol in substantial concentrations. The cytoplasmic RIG-I receptor evolved as a solution to this problem. It senses pathogen-derived RNA while ignoring the huge concentrations of host RNA molecules present in the cytosol by using two sequence-independent structural criteria. The first is that RIG-I requires a blunt-ended double-stranded RNA target with a terminal 5′ triphosphate group [20]. This provides for a large degree of specificity because in general host RNAs do not fit this bill. However, biology being what it is, this is not an absolute criterion, and host molecules with this type of structure can indeed be detected in the cytosol. The second specificity-ensuring feature is that the RIG-I binding site's structure has been selected to ensure that host mRNA molecules cannot enter it. The first nucleotides of metazoan mRNA molecules are methylated on the 2'-OH position of the ribose, and these methyl groups prevent the end of a double-stranded RNA getting past a histidine residue, which is strategically placed at the entrance to the RIG-I binding site. This particular histidine is highly conserved in RIG-I molecules all the way from sea anemones to mammals, and because of this structural barrier, the activation of RIG-I is restricted to viral RNAs, which are not methylated in this way [21]. The remarkable feature of this "self" versus "non-self" discrimination system is that it has not been selected to promote interaction with "non-self" but rather to specifically exclude interaction with "self". It is, thus, a rare example in innate immunity of a "tolerance" mechanism [21] similar in principle to the tolerance systems of adaptive immunity (Sect. 4.5.3).

3.6.4   Detecting Intracellular Bacterial Pathogens

Viruses are not the only intracellular pathogens in metazoans. The extracellular space, being full of innate immunity's soluble and cell bound antimicrobial receptors, is an unfriendly environment, and so natural selection has favoured the evolution of bacterial pathogens that can evade these dangers by hiding inside cells. These intracellular bacterial pathogens in turn drove the selection of intracellular pathogen-detecting molecules in the host, and the cytosol is generously stocked with such receptors. Prominent amongst these are the "NOD-like receptor" (NLR) family of multi-domain proteins (see Appendix B). Like the TLRs (Sect. 3.3) NLRs are a gene family that has undergone massive expansion and contraction during evolution. In the sea urchin there are 98 NLR genes, in zebra fish 400, while in humans there are only 22 [22]. Some, like the "NOD" or "NAIP" receptors, target bacterial structures that are also targeted by soluble or cell surface receptors, while others respond to a range of pathogen or stress-associated molecules. Currently the list of their ligands is both large and far from complete.

3.6.5 Detecting Endogenous "Danger" Signals

For many years it was thought that the repertoire of innate system pathogen detectors was inherently "safe" because for hundreds of millions of years the selection criterion was assumed to have been, "survival of those whose innate receptors are not directed against 'self'". However, things turn out not to be so simple, and it is now clear that the innate immune system is not restricted to fending off pathogens, but instead has been saddled with additional "housekeeping" jobs such as clearing away the corpses of apoptotic cells, removing trauma-induced tissue debris and remodelling tissues in development and metamorphosis. Because of this innate immunity must also detect so-called endogenous danger signals. These are signals which are generated in situations in which damage to host tissue occurs, irrespective of whether that damage is associated with the presence of an infectious agent or not [11]. For this purpose innate receptors have been selected that detect endogenous molecules that are normally locked up inside cells. One example is the cytoskeleton component F-actin, which is normally present inside the cytosol, and is only released when cells are lysed [23]. Once in the extracellular space it is detected as a signal of danger by the receptor DNGR-1, a C-type lectin that is expressed on the surface of certain innate sentinel cells. This is by no means a unique example, for in the meantime an increasing number of innate immune receptors are known to detect danger signals released from infected or otherwise stressed cells.

Not all danger signals are necessarily released from cells—some also work within a cell. Bacterial pathogens exploiting phagocytosis to infect a cell will enter the cytosol inside a phagocytic vacuole or endosome. To avoid the unpleasantness, which will ensue when this endosome fuses with a lysosome, random mutation and natural selection have provided *Listeria monocytogenes* with a pore-forming toxin—listeriolysin—that dissolves the endosome membrane, thus allowing the bacterium to escape into the cytosol (see Sect. 2.2.5). As the endosome starts to disintegrate, glycans on its inner surface are now exposed, and these are recognised as danger signals by cytosolic galectins, which then organise the formation of an autophagic double membrane around the damaged endosome. The autophage vesicle then fuses with a lysosome. A phagocytosed bacterium that is not quick off the mark in escaping from the disintegrating endosome, ends up inside an autophagy vesicle—a simple case of "out of the phagocytic frying pan into the autophagic fire".

These danger signals should, of course, only be present in significant amounts when there has been some deviation from the homeostatic norm, yet they will normally be present at whatever low level results from the balance between the mechanisms forming them and the mechanisms that are responsible for removing them. This balance will define a threshold level of the danger signal that innate immunity would do well to ignore. Any disturbance in this threshold-setting mechanism or any change, which raises the effective affinity of the receptor for the danger signal, will then lead to a breakdown of the innate system's ability to discriminate "danger" from "normal self". Because of this, once innate immunity learned to detect endogenous structures it lost its immunological innocence, and this made possible the development of auto-inflammatory disease [24].

3.6.6   *"Missing-Self" and the Evolution of "Natural Killer Cell" Receptors*

A special type of danger signal involves "self" structures that disappear upon infection. The detection of "missing self" is an ancient mechanism first elaborated in bacteria as a defence against viral attack. Many bacteria evolved binary DNA restriction systems composed both of a DNA methyl transferase, which methylates bases in a short defined DNA sequence, together with an endonuclease that cuts specifically only the unmethylated form of the sequence. The sequences involved are generally short, and so they are also common and will be expected to occur frequently in DNA. Since the endonuclease cannot cut the bacterial DNA at the methylated site, the bacterial genome is fully protected. However, when an unwary virus infects the cell, its sequences are missing this all-important "self" signature. They are not methylated, and so they will be detected by the endonuclease, which then trashes the viral genetic information.

"Natural killer" (NK) cells are lymphocytes participating in innate immunity, which use this sort of "missing self" strategy. They have the most unusual property that they express a suite of germline-encoded receptors directed against "self" structures and because some of these structures are present on almost all cells, natural killers can, in principle, kill any and all cells in the body. Luckily they don't do this because while roughly half of these receptors activate the killing program, the others inhibit it. The balance between the activating and inhibiting signals has been selected to be such that inhibition wins—and so the NK cells do not kill [25]. What then is the point of having them? The answer is that many of the inhibitory receptors bind the "Class-I Major Histocompatibility Complex" (MHC) molecules that are expressed on the surface of all nucleated cells (see Sect. 4.6.2). These MHC molecules play an essential role in adaptive immunity, for without them the adaptive system is unable to "see", and then destroy, virus-infected cells. Since host–pathogen interactions take the form of a perpetual arms race, it is no surprise that many viruses try to make the cells they have infected invisible, by down-regulating the expression of these MHC molecules. This is where the NK cells come to the rescue of the host, for if an NK cell meets a virus-infected cell that no longer expresses MHC molecules on the surface, then the inhibitory signals are gone, but many of the activating signals are unchanged. The balance of signalling is decisively shifted from "inhibition" to "activation", and the NK cell will destroy the virus-infected cell. The same is true for the NK cell response to those tumour cells that down-regulate MHC expression. We will return to these NK cells and their mode of action later in Sect. 3.8.1, but for the moment we should look briefly at two further remarkable aspect of the evolution of these receptors that permit the cell's response to missing self.

The first point is that in humans many members of the NK cell receptor repertoire belong to a family of molecules whose extracellular ligand-binding part is composed of IgSF domains (see Appendix B). In the mouse, the functionally equivalent receptors have a totally unrelated structure, for their extracellular ligand-binding part is composed instead of a C-type lectin domain. Thus, though the molecular form of the mouse receptors is radically different from that of the human ones, nevertheless these two repertoires are functionally interchangeable and have been readily swapped during the course of mammalian evolution [26].

The second remarkable aspect of the evolution of these two families of receptors has to do with the ways in which they transmit signals into the NK cells. Remember that both the IgSF and the C-type lectin NK receptor families contain both activating and inhibitory members. The essential difference between an activating and an inhibitory receptor lies in the way that they are linked to the NK cell's signal transduction machinery. The inhibitory receptors of both families possess a small cytoplasmic domain of around 45 amino acids, which initiates the inhibitory signal transduction cascade. In contrast, the activating receptors of both families have instead a transmembrane domain with an unusual structure, which allows it to interact in the cell membrane with an adaptor protein whose intracellular part then activates the downstream activating signal transduction pathway. Thus, the all-important means of connecting the extracellular domains either to activating or to inhibitory intracellular signal transduction systems has been retained—even though the extracellular domains have been switched from C-type lectin in the mouse to IgSF in humans.

Bear these points in mind, for an even more dramatic case of retention of function despite a switch in receptor structure will be coming up in Chap. 4 when we discuss the different adaptive system receptors used by non-jawed and by jawed vertebrates.

3.7   Living with Commensals

Ligands like LPS, peptidoglycan or flagellin, which are only produced by bacteria, would seem to be almost ideal targets for innate immunity, yet receptors directed against them will also detect the large number of microorganisms that are part of our "microbiome" of bacterial symbionts. For example, the intestine contains enormous numbers of both Gram-positive and Gram-negative commensal bacteria, and these are beneficial, partly because some are able to breakdown components of the food for which the host has no suitable hydrolytic enzymes, and partly because, by occupying the intestine, they deny space to pathogens.

If sentinel cells were to be constantly alerted by these intestinal bacteria, then the innate immune system would be chronically activated and that would have devastating consequences for the host. How is chronic activation to be avoided? This is one of those situations where evolution has had to find a compromise solution. The cells forming the intestinal surface must defend themselves against attack by bacteria from the lumen, which otherwise would rapidly invade the body. Yet at the same time this surface cannot be converted into an impenetrable barrier, for its primary function is to act as the gateway through which food is absorbed. The compromise solution that has been selected is to largely deny microbes access to the luminal surface of the intestine by covering it with a layer of mucous, which contains various antibacterial peptides and IgA antibodies. Bacteria outside of this mucous layer need not be considered an obvious danger, while those few, which have found a means to penetrate into the mucous and approach the intestinal epithelium, may need to be vigorously opposed. Nevertheless, the situation is far from black and white, for it is now clear that the immune system shapes the microbiota present in the intestine, and that equally the microbiota contributes to shaping the immune system. The nature of these interactions and their consequences are currently one of the hottest topics in immunology [27].

3.8 Beyond Receptor-Ligand Interactions: Immune Systems as Computational Devices

There is a lot more to immune defence than the mere availability of a repertoire of receptors that can be activated by a broad array of microbial ligands. The real key to innate immunity's success has to do with the way in which the interaction of its receptors with their ligands is interpreted. A simple set of knee-jerk responses to the detection of microbial ligands is not going to be of much value in a world where new pathogens arise all the time, while old ones develop ever more sophisticated virulence mechanisms. The immune system is thus constantly facing new challenges, and to be of continuing selective value it must routinely deal with novel and totally unexpected problems—and these cannot be solved with simple monotonous reflexes. The ability to quickly find the optimal response to the unexpected requires more than just the information that this or that microbial ligand is now in close proximity to some sentinel cell. Instead, information from many different receptors and from many different cells has to be collected and integrated so that the optimal response can be put together. In this light, the immune system can be regarded, above all else, as something that evolved into a computational device.

3.8.1 Signal Transduction

Innate immunity's cell bound receptors report their interaction with their ligand into the cell's signal transduction system. Signal transduction involves the chain of events whereby a ligand–receptor interaction leads to some change in cell physiology. It may be a change in the activation state of cellular enzymes, as happens, for example, in receptor triggered phagocytosis, or it may involve a change in the cell's pattern of gene transcription. We can take the signalling events resulting from the interaction of a TLR at the cell surface with its ligand, as an example of this latter

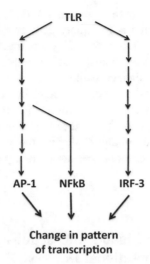

Fig. 3.7 A simplified scheme illustrating the signal transduction pathway from a cell surface TLR. Each arrow represents one step in the pathway. Information flow is defined by the flux of modification (phosphorylation, ubiquitinylation, sumolation, etc.) of the intermediates. The information flow in the pathway can be modulated by cross-talk from other signal transduction pathways. TLR signalling may end in the activation of the transcription factors AP-1, NFkB and IRF-3, which are able to stimulate the transcription of genes including those encoding pro-inflammatory cytokines and type 1 interferons

case. A simplified scheme of the signalling transduction pathway is shown in Fig. 3.7. In brief, after ligand binding the TLR's intracellular domain interacts with adaptor molecules, which lead on to a chain of modifications of signal transduction intermediates that ends up with the activation of the transcription factors NFkB, AP-1 and IRF-3. These transcription factors then promote changes in the pattern of gene transcription within the cell. Those wishing for the details of the protein modification reactions along the way will find them in any immunology textbook. For us the important point is merely that there are a large number of intermediates involved. Why this seemingly unnecessary complexity?

The large number of intermediate steps is there to permit innate immunity to operate as a computational device. At the simplest level this works because the sentinel cells of innate immunity express not just one, but numerous different cell surface and intracellular receptors—TLRs, C-type lectins, NOD-Like Receptors, Fc-receptors, as well as receptors for cytokines, chemokines, complement factors, and more. All of these provide the cell with information about its local environment, and thus inform it about any departures from the homeostatic norm. Different receptors engage different signal transduction pathways. However, the information from the different activated receptors does not simply flow down parallel signal transduction pathways, for these different pathways are not strictly segregated. They share common components and choke points, so that the flux of signal strength through one pathway will be altered and modulated by the activity of other receptors

[28]. In addition, the information flow through the signal transduction system will also be modulated by the signals entering it as the cell constantly monitors its own physiological status. This extensive "cross-talk" between signal transduction pathways provides for an integrated and finely nuanced response to a challenge.

Little is known about how cell autonomous computation based on changing information fluxes in signal transduction pathways can give cells the ability to learn from the past, and so determine their assessment of the present, but these are features that are important for an immune system. Their importance is particularly obvious in the case of the NK cells described in Sect. 3.6.6. These cells make decisions as to whether to kill other cells on the basis of the information flowing into their signal transduction network from receptors that they express on the surface. As described above, some of these receptors send activating signals into the NK cell, while others send inhibitory signals. It is the balance between activating and inhibitory signals that determines whether an NK cell will kill a potential target or not. How is this life-or-death-determining balance to be established? One thing is clear—it is not something that can be imposed by hard-wired genetic mechanisms. The reasons for this are straightforward. Each mature NK cell expresses only a subset of all the activating and inhibiting receptors encoded in its genome. The choice of which receptors are to be expressed appears to be simply a matter of chance. The result is that within one individual, different NK cells express different combinations of receptors. Each developing NK cell therefore has to adjust the balance of activating and inhibiting signals to the spectrum of receptors that it expresses. Furthermore, many of the receptors are polymorphic, as are their ligands, and the genes encoding these receptors and ligands assort independently at meiosis. Worse still, the different polymorphic forms of a particular ligand may bind with very different affinities to their cognate receptor. The result is that there is no way of predicting which mix of receptors will be expressed in any given developing NK cell, and no way of predicting what signal flux will be generated by a receptor–ligand interaction. There is also no guarantee that a cognate ligand will be available for each of the receptors expressed [29]. Because of all this each developing NK cell has to establish the necessary balance between activating and inhibitory signals on its own—in other words, it has to program its signal transduction computation network itself.

In the immune system, cell autonomous information processing is only the first part of the computation of an optimal response to changes in the homeostatic norm. The different immune cell populations express different mixes of receptors, so that what cell A makes of the information available in its environment may be very different from what cell B makes of the same information. Just as the different receptors on one cell interact with each other by cross-talk in the signal transduction cascades, so the different innate cells share their assessments of the current situation by interacting via the cell surface molecules they express and by the cytokines they release. Nevertheless, this is not a "Tower of Babel" situation where everybody talks loudly, nobody listens and nothing gets done; for neither do all cells express all cytokines or all cytokine receptors, nor do these different cell populations express the same arrays of cell interaction molecules on their surfaces. Because of this, though the sharing of information between the cells is extensive, it is nevertheless neither random nor unstructured. The ability of the various cell populations to share their assessments of the current situation and to make their intentions clear to the others is what immunity needs to mount a judiciously chosen, optimal response. Only in extreme cases, such as during the "cytokine storm" associated with sepsis, does this mechanism for information sharing become overloaded and the innate immune system's capacity for computation crashes.

Computational devices are in the business of making decisions. The receptors of innate immunity serve to provide the input data, but there is a long way from there to a final output, and many aspects of the ways in which a response develops, is regulated, and then is terminated, remain black boxes.

3.9 The Outputs

Receptors that will detect potential pathogens, and signal transduction pathways that will pass that information along, are themselves not enough—there has to be some means of destruction waiting at the end of the chain. The options available are all very ancient. Phagocytosis followed by fusion of the endosomes with lysosomes is both a means of sustenance and a means of defence for the single cell organisms that existed prior to the evolution of metazoans. In multicellular organisms the detection of the presence of a pathogen results in the elaboration of an inflammatory response, which is essential to rapidly recruit the phagocytes that will deal with the infection. The evolution of the means to attract the effector cells required the prior development of the endothelial-lined circulatory system that evolved early in vertebrate development.

3.9.1 Cell Movement and the Inflammatory Response

Bringing phagocytes—principally neutrophil granulocytes—to a site of infection requires the ability to direct the movement of cells within an organism, and that is, in principle, not a problem, for it is a skill that is as old as metazoan life. Cell movement is central to the formation of defined morphological forms, to the re-organisation of cells and tissues during embryogenesis, and to the movement of cells during tissue repair or remodelling. It is dependent on the ability of a cell to interact with its neighbours, and that requires some form of cell-to-cell interaction via cell adhesion molecules. Some of these cell adhesion molecules—such as cadherins and integrins—are present already in single cell pre-metazoans [30]. In metazoans, numerous families of cell adhesion molecules permit reversible interactions between cells or between cells and the extracellular matrix, while gradients of soluble factors or mechanical forces provide information as to where the cell is supposed to be going. Cells can thus find their way to their proper location by crawling over other cells or over the extracellular matrix.

Immunity, however, faces problems of cell movement, which are altogether different in scale. Neutrophils and other immune cells must be brought quickly to within striking distance of the site of an incipient infection, and in a large animal like a human being, or an elephant, it would take an age for these cells to crawl from their birth place in the bone marrow to an infection at some distant part of the body. To solve this problem immunity borrows "inventions" that were selected for completely different purposes. One of these involves the circulatory systems that permitted the evolution of animals larger than just a handful of cells. Tiny animals can ensure that all of their cells have access to the oxygen and nutrients they need by simple diffusion. As an animal gets larger this becomes ever more difficult, and so increasing size becomes dependent on some form of circulatory system, and these come in various forms. In "open" systems, as for example in insects, a heart will pump fluid around the body cavity to bathe the internal organs. In "closed" systems, as for example in annelid worms, the blood is pumped through vessels, and this allows for a much more efficient distribution. In most animals with a closed circulation—as in the earthworms or in the cephalochordate *Amphioxus*—the vessel is formed of extracellular matrix. As the matrix material is somewhat permeable, the pressure, and hence the rate at which the blood can be circulated, is limited. This problem was solved early during vertebrate evolution by the invention of the continuous endothelial lining of the vessels. These endothelial cells form a tight impermeable sheath between the blood and the underlying matrix, and this permits the use of higher blood pressure, which increases the rate of transport of nutrients. With their smooth surfaces, endothelial cells also reduce drag that otherwise impedes flow [31].

The invention of the endothelium was also the key step in the evolution of an efficient system of systemic immunity, for—unlike the inanimate extracellular matrix-these endothelial cells form an "intelligent" surface that can be rapidly and precisely induced to respond to signals emanating from the underlying tissue. Vessel endothelial cells respond to signals such as IL-1β or other "pro-inflammatory cytokines" from the adjacent infected tissue, by changing the expression of cell surface molecules. These activated endothelial cells somewhat reduce their contacts with their neighbours, so that solute molecules in the blood, such as agglutinins, antibodies, complement components and other factors, can now leak out into the infected tissue. Activated endothelial cells also engage in a finely tuned set of interactions with circulating immune cells, to permit neutrophils, monocytes or lymphocytes to leave the bloodstream close to the site of an infection. The way that this is done in vertebrates required three sets of co-evolving molecules. The first is composed of the selectins and selectin ligands, the second of "chemotactic cytokines" or "chemokines" and their receptors, and the third is composed of integrin and integrin ligand pairs.

The system works more or less as follows. A cell leaving the human heart will be moving in the bloodstream with a velocity of around 40 to 50 cm/s. It is hard to envisage a means of persuading a cell moving that quickly to leave the circulation safely. In the capillaries, however, that same cell will be travelling roughly three orders of magnitude more slowly. Indeed, the large neutrophils have to squeeze their way through the capillaries, and this causes something of a traffic jam as the smaller, more numerous and faster moving erythrocytes pile up behind. But as the neutrophil emerges into the post capillary venule the erythrocytes can now accelerate past it, pushing the larger cell towards the wall as they go by. It is usually in these "post capillary venules" that the cells leave the circulation. The problem is that were a cell accelerating into a post capillary venule to be suddenly locked onto the endothelial surface, then shear force would rip it apart. It first has to be slowed down. This is where the selectins and the selectin ligands come into their own. Selectins are unusual cell adhesion molecules because, though they interact quickly with their substrate, they also do not hold on for long. Because of this a rapid series of hold-release cycles ensues, and this progressively slows down the leukocyte as it rolls across the surface of the endothelium. Selectin–selectin ligand interactions act as the leukocyte version of the anti-block system in a car's brake.

The blood neutrophil has now been slowed down by selectin–selectin ligand interactions to such an extent that it can be firmly attached to the vessel wall, but this requires a tight interaction between members of two classes of cell adhesion molecules—the integrins expressed on the neutrophil and the integrin ligands on the endothelium. There is no time for these molecules to be produced by transcription and translation, so both sets are constitutively expressed. However, the integrins on the neutrophil surface are present in an inactive conformation so that they cannot normally bind to their ligands on the endothelium. This inactive integrin conformation can be quickly changed to an active form, but this requires a second interaction in which the activated endothelium signals the identity of the cell type that is to be encouraged to exit from the circulation. This it does by secreting chemokines. Were these small protein molecules merely to be secreted in the normal way, then they would immediately be swept away into the bloodstream, and nothing would be achieved. However, the secreted chemokines bind tightly to the extracellular matrix on the endothelial cell, and thus stay in place as markers of a local infection. There are a large number of different chemokines and chemokine receptors, so that only when the leukocyte expresses the appropriate complementary receptor for the chemokine bound on the surface of the endothelium will this second interaction be successful, and lead to the activation of the neutrophil integrins. Only if the leucocyte and the endothelium express a matching integrin and integrin ligand pair will this third step take place, and finally lock the cell firmly to the endothelial surface. Neutrophils adhered on the surface of the endothelium must now make their way between, or in some cases even through, the endothelial cells. This passage is not nearly as well understood as the initial interaction with the endothelial surface. It is, however, rapid, is actively supported by both partners and it does not lead to a loss of the endothelial barrier function.

To exit from the blood a cell must therefore express the molecules that permit it to engage with the endothelium in the three successive interactions: selectin–selectin ligand, chemokine–chemokine receptor and finally integrin–integrin ligand. These three successive interactions have been described as forming a "postal code" that defines where a particular leucocyte may exit from the blood [32]. The system is dynamic, since the signals on the endothelial cells can be changed so that different types of leukocytes can be recruited at different stages of the response: in general neutrophils first, then monocytes and finally lymphocytes. This postal code system is an astonishing example of the co-evolution of families of receptors and their ligands. It is central to the recruitment of neutrophils to sites of inflammation, but it has a much broader significance, for it is also used in quite different contexts to regulate the trafficking of all blood leucocytes, and the same principles regulate the movement of immune cells through lymphoid tissues.

Once beyond the endothelial barrier the neutrophil must find a way through the extracellular matrix below, and then follow chemokine gradients towards the activated sentinel cells in the infected tissue. Having reached the infected tissue, neutrophils get to work. They rapidly phagocytose pathogens and destroy them with reactive oxygen intermediates and proteases. Those that they cannot eat, they engulf in "Neutrophil Extracellular Traps"—in short the whole panoply of killing skills developed by pre-metazoans.

3.9.2 Complement-Mediated Phagocytosis

The neutrophils attracted to a site of infection are there to destroy the pathogen by phagocytosis, and the mechanism of phagocytosis remains basically what it always was. Yet over the course of time it has benefited from numerous technical improvements, and the greatest of these came with the evolution of the complement system. Complement provides a means of labelling particles or molecules that have been identified as "non-self" either by innate immune receptors, such as MBL, or by antibodies.

The central actor in the complement system is a molecule called "complement factor 3" (C3), which contains hidden within it a thio-ester bond that on exposure to an aqueous environment becomes highly reactive. The evolution of the complement system is a classical tale of adapting a good idea—the invention of this thio-ester bond—to the ever changing needs of phylogeny. The origin of metazoan thio-ester proteins is buried in a chicken and egg argument in which a protease inhibitor called α2 macroglobulin plays the part of the chicken and C3 that of the egg. These are the two earliest known members of the thio-ester protein (TEP) family. Both are present in Cnidarians but not in the choanoflagellate *Monosiga brevicollis* or the sponge *Amphimedon queenslandica* (see Appendix A), suggesting that this TEP family arose early in eumetazoan evolution [33]. What the TEP does in these basal metazoans is not known.

During the course of evolution TEP gene numbers expanded and contracted differently in different groups of animals. The bony fish have massively expanded their C3 genes, while in mammals the two rounds of whole genome duplication early in vertebrate evolution provided us with three copies of the C3 type thio-ester protein gene, C3, C4 and C5—though C5 has in the meantime lost its reactive thio-ester bond. Insects somehow lost their C3 gene, but they "re-invented" it from the alpha-2 macroglobulin sequence, which they had retained—a good example of strong selection pressure resulting in analogous solutions. Thus, *Drosophila melanogaster* currently has six thio-ester containing alpha-2 macroglobulin-related genes. One is a pseudo gene, and a second has lost the sequences needed to form the thio-ester bond, but the others show C3 like activity—at least in vitro.

In vertebrates, complement factor 3 leads two quite different lives. In its first it is an abundant but reserved, and unsociable serum protein, which basically does nothing at all. In fact C3's most striking characteristic is that, apart from a handful of proteases, it has no interaction partners. On activation, however, it changes in the twinkling of an eye into a wildly gregarious molecule with so many specific interaction partners that there isn't space on its surface to accommodate binding sites for them all. C3 solves this problem by acting as a molecular strip tease artist, gaily abandoning domains to expose ever new interaction surfaces, and in the process generating a plethora of fragments—C3a, C3b, C3c, C3dg, C3d, C3f and C3g—all of which have their own peculiar biological activities.

Activation of C3 into C3b by "C3 convertases" results in the thio-ester bond being exposed on the surface where it reacts with almost anything in the immediate vicinity. Indeed C3b very quickly finds a suitable partner, for the half-life of the thio-ester bond on the surface is thought to be around one ten thousandth of a second.

3.9.2.1 C3 Convertases

There are a number of ways to activate C3 (Fig. 3.8). The so-called lectin pathway (Fig. 3.8a) is initiated by innate immune receptors that have identified and bound to a "non-self" surface. The best-studied example involves the vertebrate MBL (Fig. 3.5a). This lectin is associated with two "Mannose-Binding Lectin Associated Serine Proteases" (MASPs), which are activated when MBL binds to a suitable mannose rich microbial surface. These then lead to the formation of the "lectin pathway C3 convertase" which is covalently linked to the surface to which the MBL had bound. By making the interaction of soluble innate receptors with a "non-self" structure a prerequisite for the formation of the convertase, the production of the phagocytosis-inducing C3b should be taking place only in direct vicinity of the "non-self" structure. MBL is restricted to vertebrates, but tunicates and other invertebrates have lectins that, together with their MASP homologs, can form C3 convertases.

With the appearance of protein-based adaptive immunity in jawed vertebrates (see Chap. 4), the complement system evolved the ability to recognise antibodies that have bound their antigens. In this so-called classical pathway of complement activation (Fig. 3.8b) a molecule called complement factor 1q (C1q) binds to antibody–antigen complexes. C1q has a structure that is reminiscent of MBL and it is associated with two serine proteases that are homologous to the MASPs. When the C1q binds to an antibody-antigen complex, these MASP-like associated proteases generate the "classical pathway convertase", which converts C3 into C3b. As in the lectin pathway the fact that the convertase (C4bC2b) is covalently linked to the "non-self" structure, together with the very short half-life of the C3's thio-ester bond,

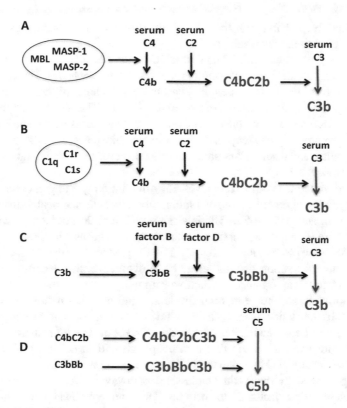

Fig. 3.8 C3 and C5 convertase pathways. Complement factor C3 contains an intra-molecular thio-ester bond buried in a hydrophobic pocket. C3 remains inactive for as long as the three-dimensional structure of the protein keeps this bond sequestered. This sequestration is ensured by one of C3's structural domains, which is so placed that it locks the molecule in this inactive conformation. Activation requires that this domain no longer perform this blocking function. This can be achieved if the blocking domain is physically removed by one of several activating proteases known as "C3 convertases". C3 convertases chop the molecule into two parts: the larger C3b fragment that contains the all-important thio-ester bond, and the smaller C3a fragment. As soon as the C3a domain is removed, the C3b fragment undergoes a dramatic change of shape during which the high-energy bond swings out onto the surface, where it will react immediately with any hydroxyl or amino groups in the vicinity [34]. The formation of the various convertases is indicated in the figure. (a) The lectin pathway convertases: the Mannose-binding lectin, (MBL) is associated with two inactive proteases—MASP-1 and MASP-2. When the lectin binds its target these proteases are activated and cut the serum protein C4 into two fragments—C4a and C4b. C4 is a paralog of C3 and also contains a thio-ester bond, which is present on the surface of the C4b fragment. This highly reactive bond will interact with hydroxyl or amino groups on the surface to which the lectin had absorbed. C4b has proteolytic activity, which permits it to convert the serum protein C2 into the active C2b fragment. This associates with C4b to form the lectin pathway convertase C4bC2b that is able to convert C3 to the active C3b form. (b) The classical pathway convertases. The classical pathway is initiated when complement factor C1q binds to the Fc region of an antibody molecule that has bound its antigen. C1q is associated with two inactive proteases that are homologous to the MASPs. These proteases—C1r and C1s—are activated on binding of C1q to its target and lead to the formation of the C4bC2b convertase. (c) The alternative pathway convertases. The activated

makes it highly probable that the C3, which this convertase generates, will also covalently bind to the "non-self" surface.

There are two results of this. The first is that microbes identified by lectins or by antibodies can be quickly and efficiently removed by macrophages. The value of this can be seen from the fact that people who lack components of these pathways are liable to repeated incidents of severe bacterial infection. The second feature is that it leads to the decoration of soluble antibody–antigen aggregates with C3b, and to their subsequent removal by phagocytes. If this does not take place, then these complexes may accumulate and result in "immune complex disease" by clogging up the kidneys and other organs.

In jawed vertebrates, C1q recognises not only antibody–antigen complexes but also a wide range of other ligands, including apoptotic cells and pentraxins that have bound their ligands (Fig. 3.5b). The jawless vertebrates do not have immunoglobulins, but there is nevertheless a C1q homolog, which acts as a lectin. In association with a MASP-like protein, it activates C3. The C1q receptor of the classical pathway of complement activation may therefore have evolved from a lectin-type receptor present prior to the appearance of adaptive immunity.

The enormous advantage of both the lectin and classical pathways is that the lectin or C1q must first identify a "non-self" structure so that formation of the convertase, and hence activation of C3, takes place in the immediate vicinity of the target. However, to ensure that a macrophage will detect this C3b decorated structure the density of C3b on the target has to be maximised. This is achieved when the C3b generated by the lectin or classical pathway convertase interacts with a serum protein called "factor B" to form the "alternative pathway convertase". This alternative pathway is, in essence, a powerful amplification loop that allows for the production of ever more C3b. It is thought that this "alternative" pathway forms 80% or more of the C3b generated at sites at which complement activation has been initiated by the lectin and classical pathways [35].

Making the formation of the convertase dependent on the prior binding of a lectin or of C1q to the target, directs C3b generation to the vicinity of a microbial surface and certainly helps to reduce the risk that complement will attack "self" tissues, but it is not enough on its own. Apart from anything else there is a convertase-independent means of activating complement. This happens when C3 undergoes a spontaneous change of shape that exposes its thio-ester bond on the surface. This spatial displacement is energetically highly unfavourable, but it does happen, now and again, and results in the highly reactive conformational isomer called "$C3(H_2O)$". Given that C3 is present in human serum at a concentration of around 1.2 mg/ml, there is a continuous production of low levels of this activated $C3(H_2O)$ form, which like C3b

---

Fig. 3.8 (continued) C3b may interact with the serum protein called "factor B". The C3bB so formed then interacts with a serum protease called "factor D" to form the alternative pathway convertase "C3bBb". (d) The complement factor C5 convertases. Excess C3b can be recruited to form a C4bC2bC3b or a C3bBbC3b complex that act as "Complement factor 5 convertases"

can bind factor B and D to form the alternative pathway convertase (Fig. 3.8c). This time, however, the convertase is being generated free in the circulation, and it generates C3b, which in turn interacts with factor B to form yet more C3b. Activated complement is now being generated without any lectin or antibody to direct it to a microbial target. C3b generated in this way is something of a rogue elephant, for it is likely to attack and destroy host tissues, and this must be actively discouraged. For this reason, a large fraction of the many molecules involved in the vertebrate complement system are there either to inhibit the attack of activated forms of C3 on "self" structures, or to promote their preferential interaction with "non-self" surfaces.

The alternative pathway of complement activation may be very ancient, for two of the key components—C3-like and factor B-like molecules—are found already in Cnidarians (see Appendix A). However, genes for factor D have so far not been detected, nor have genes for the numerous factors required to keep the alternative pathway under control. Perhaps Cnidarians have some other way of activating factor B, and of controlling C3b. Alternatively, the Cnidarian C3-like TEP homolog may play some role unrelated to defence. Clearly there are parts of the complement story about which we currently know almost nothing.

3.9.2.2 The Membrane Attack Complex

One further twist to the complement story also accompanied the evolution of vertebrates. This involves the formation of a so-called Membrane Attack Complex (MAC) that can lead to the destruction of cell membranes. This complex is formed when sufficient C3 convertase is deposited on a surface so that a new player—"the C5 convertase"—comes on stage (Fig. 3.8d). This convertase activates C5, which in turn organises the assembly of a slew of factors (C6, C7, C8 and C9) into a complex that forms pores in cell membranes. Once the integrity of the membrane is destroyed in this way, the cell will die. The mystery surrounding this complex pathway is that it is unclear what selective pressure drove its evolution, for it is not clear what the MAC is really good for. Gram-positive bacteria are immune to it, because their membrane is protected by the thick cell wall. Gram-negative bacteria and parasites might be thought of as better targets, but people who are unable to form the MAC because they have lost one or other of the complement components C6 to C9 seem to differ from their normal siblings only in their susceptibility to infection by *Neisseria* species, particularly by *Neisseria meningitidis*. The evolution of the MAC would seem to be a rather extravagant solution to the problem of meningo-coccal infections—but one should never try to second-guess evolution.

There is no functional evidence for a MAC being formed in invertebrates, but it is formed in the agnathan lamprey. That is not just odd, it is astonishing, for lampreys are basal vertebrates that do not make antibodies, and they have no homologs of the genes for C6, C7, C8 or C9 in their genome [36]. How does the lamprey do it? The answer is that these agnathans do have an adaptive immune system (see Chap. 4). It is not based on antibodies, but rather on molecules called VLRBs, and the lamprey's ability to form a MAC requires the presence of both VLRB and C3 [36]. The retention of the ability to generate a MAC across the great divide that stands between the lamprey and jawed vertebrate adaptive immune systems is a remarkable tribute to evolution's ability to successfully tinker with things. It is also a strong argument in favour of a pivotal role of this MAC pathway—even if at the moment we cannot quite see what this might be.

3.9.2.3 Other Functions of C3

The list of the physiological systems that activated C3 participates in is long, and it seems to grow longer by the day. Both in vertebrates and in invertebrates it has a well-attested role in promoting the phagocytosis of pathogens, while in jawed vertebrates it is also crucial for the removal of antigen–antibody complexes from the circulation. However, complement also acts in a wide range of other physiological processes. It is involved in defence reactions at sites of tissue injury, where proteases of the coagulation system such as plasmin and plasma kallikrein can generate the C3b fragment in a so-called extrinsic pathway of complement activation. It is also utilised to link the innate and adaptive immune systems, since the C3d fragment bound to an antigen massively increases that antigen's ability to activate B-lymphocytes [37]. Furthermore, the C3a and C5a fragments released during complement activation are not just pieces of molecular junk, for they play important roles as accelerators of the development of an inflammatory response. They locally increase the permeability of the endothelium, and act as attractants and activators of neutrophils, basophils, mast cells, eosinophils and macrophages. Beyond that, C3 activation is involved in a range of other activities that are not related to innate defence, including pruning the synapses formed between neurones [38] and in regulating intracellular metabolism [39].

3.9.3 Programmed Cell Death

Cell death may be brought about in a number of different ways. It can result from a random accumulation of damage, which finally causes the membrane to rupture and the cytosolic contents to pour out. This form of death is classically known as "necrosis". Alternatively a cell may be persuaded to induce a program of events that lead to the activation of proteases called caspases, which destroy essential components, and thus kill it. The cell dies without rupture of the membrane and is phagocytosed by macrophages. This is the classical picture of programmed cell death, referred to as "apoptosis". In the meantime it is clear that programmed death comes in a variety of forms of which necrosis and apoptosis may represent the ends of a spectrum [40]. The form of cell death that operates as a last resort defence against pathogens in the cytosol—"pyroptosis"—is one that is halfway between necrosis and apoptosis. It resembles apoptosis in that it is caspase dependent, but death is accompanied by rupture of the membrane, as in necrosis. Elegant arguments have been put forward to support the notion that pyroptosis is perhaps the "original" form of programmed cell death, and that its invention preceded the evolution of metazoans [41]. Pyroptosis is initiated when a caspase is activated by binding to some repetitive structure in the cytosol. This repetitive structure may be Gram-negative bacterial LPS, which binds and activates human caspases 4 and 5, which in turn can then activate downstream effector caspases that kill the cell. Alternatively, NOD-Like Receptors may oligomerise to form structures known as an inflammasomes. Caspase-1 is activated by recruitment into the inflammasome and may then cleave a protein called Gasdermin-D, which in ways that are not quite clear at the moment, initiates pyroptotic cell death. Merely killing an infected cell would not necessarily be a good idea, because the infectious contents would then be released. However, the activated caspase-1 cleaves not only gasdermin-D but also the inactive form of the cytokine IL-1β, converting it into the active form. Active IL-1ß released from the pyroptotic cell serves to start the inflammatory program that attracts neutrophil granulocytes to the site of infection, where they do their very best to gobble up the dying cell and all of its contents. Perhaps this explains why many viruses target key steps in the cell death programs.

3.10 Who Needs More?

The innate immune system is equipped with a repertoire of receptors, which—honed over millions of years—have worked in the past to detect the presence of pathogens. Activation of endothelial cells in the local capillaries allows antimicrobial molecules and immune effector cells to be concentrated at a site of infection where they will aggressively attack the invading pathogens. Who needs more? The answer is that innate immunity suffers from the weakness that many pathogens, with their short generation times, have the opportunity to evolve new means of outwitting our limited repertoire of innate receptors very much faster than we can gain new receptors by the fixation of suitable mutations. This "generation gap" (see Sect. 1.10) sets a limit on the effectiveness of defence systems that operate with receptors encoded in the germline. Natural selection of alternative immune repertoires that are not limited in this way would have provided a decisive selective advantage. The solutions that emerged are known as adaptive immune systems and will be considered in the next chapter.

References

1. Andra J, Herbst R, Leippe M (2003) Amoebapores, archaic effector peptides of protozoan origin, are discharged into phagosomes and kill bacteria by permeabilizing their membranes. Dev Comp Immunol 27(4):291–304
2. de Koning AP et al (2011) Repetitive elements may comprise over two-thirds of the human genome. PLoS Genet 7(12):e1002384
3. Leulier F, Lemaitre B (2008) Toll-like receptors—taking an evolutionary approach. Nat Rev Genet 9(3):165–178
4. Dickson KA, Haigis MC, Raines RT (2005) Ribonuclease inhibitor: structure and function. Prog Nucleic Acid Res Mol Biol 80:349–374
5. Parthier C et al (2014) Structure of the Toll-Spatzle complex, a molecular hub in Drosophila development and innate immunity. Proc Natl Acad Sci U S A 111(17):6281–6286
6. Zhang Z et al (2016) Structural analysis reveals that toll-like receptor 7 is a dual receptor for guanosine and single-stranded RNA. Immunity 45(4):737–748
7. Liu M, Grigoriev A (2004) Protein domains correlate strongly with exons in multiple eukaryotic genomes--evidence of exon shuffling? Trends Genet 20(9):399–403
8. Yang X et al (2016) Widespread expansion of protein interaction capabilities by alternative splicing. Cell 164(4):805–817
9. Racimo F et al (2015) Evidence for archaic adaptive introgression in humans. Nat Rev Genet 16 (6):359–371
10. Dannemann M, Andres AM, Kelso J (2016) Introgression of neandertal- and denisovan-like haplotypes contributes to adaptive variation in human toll-like receptors. Am J Hum Genet 98 (1):22–33
11. Matzinger P (2002) The danger model: a renewed sense of self. Science 296(5566):301–305
12. de Zoete MR et al (2011) Cleavage and activation of a Toll-like receptor by microbial proteases. Proc Natl Acad Sci U S A 108(12):4968–4973
13. Boyd AC et al (2012) TLR15 is unique to avian and reptilian lineages and recognizes a yeast-derived agonist. J Immunol 189(10):4930–4938
14. Willems L, Gillet NA (2015) APOBEC3 interference during replication of viral genomes. Viruses 7(6):2999–3018
15. Conticello SG et al (2005) Evolution of the AID/APOBEC family of polynucleotide (deoxy) cytidine deaminases. Mol Biol Evol 22(2):367–377
16. Kranzusch PJ et al (2015) Ancient origin of cGAS-STING reveals mechanism of universal 2',3' cGAMP signaling. Mol Cell 59(6):891–903
17. Martin M et al (2018) Analysis of drosophila STING reveals an evolutionarily conserved antimicrobial function. Cell Rep 23(12):3537–3550 e6
18. Goto A et al (2018) The kinase IKKbeta regulates a STING- and NF-kappaB-dependent antiviral response pathway in drosophila. Immunity 49(2):225–234 e4
19. Stetson DB et al (2008) Trex1 prevents cell-intrinsic initiation of autoimmunity. Cell 134 (4):587–598
20. Rehwinkel J et al (2010) RIG-I detects viral genomic RNA during negative-strand RNA virus infection. Cell 140(3):397–408
21. Schuberth-Wagner C et al (2015) A conserved histidine in the RNA sensor RIG-I controls immune tolerance to N1-2'O-methylated self RNA. Immunity 43(1):41–51
22. Howe K et al (2016) Structure and evolutionary history of a large family of NLR proteins in the zebrafish. Open Biol 6(4):160009
23. Sancho D, Reis e Sousa C (2013) Sensing of cell death by myeloid C-type lectin receptors. Curr Opin Immunol 25(1):46–52
24. Roers A, Hiller B, Hornung V (2016) Recognition of endogenous nucleic acids by the innate immune system. Immunity 44(4):739–754
25. Daeron M et al (2008) Immunoreceptor tyrosine-based inhibition motifs: a quest in the past and future. Immunol Rev 224:11–43

26. Carrillo-Bustamante P, Kesmir C, de Boer RJ (2016) The evolution of natural killer cell receptors. Immunogenetics 68(1):3–18
27. Hooper LV, Littman DR, Macpherson AJ (2012) Interactions between the microbiota and the immune system. Science 336(6086):1268–1273
28. Ostrop J, Lang R (2017) Contact, collaboration, and conflict: signal integration of Syk-coupled C-type lectin receptors. J Immunol 198(4):1403–1414
29. Long EO et al (2013) Controlling natural killer cell responses: integration of signals for activation and inhibition. Annu Rev Immunol 31:227–258
30. Sebe-Pedros A, Degnan BM, Ruiz-Trillo I (2017) The origin of Metazoa: a unicellular perspective. Nat Rev Genet 18(8):498–512
31. Monahan-Earley R, Dvorak AM, Aird WC (2013) Evolutionary origins of the blood vascular system and endothelium. J Thromb Haemost 11(Suppl 1):46–66
32. Springer TA (1995) Traffic signals on endothelium for lymphocyte recirculation and leukocyte emigration. Annu Rev Physiol 57:827–872
33. Nonaka M, Kimura A (2006) Genomic view of the evolution of the complement system. Immunogenetics 58(9):701–713
34. Ricklin D et al (2016) Complement component C3 – The "Swiss Army Knife" of innate immunity and host defense. Immunol Rev 274(1):33–58
35. Ricklin D, Lambris JD (2016) Therapeutic control of complement activation at the level of the central component C3. Immunobiology 221(6):740–746
36. Wu F et al (2017) A pore-forming protein implements VLR-activated complement cytotoxicity in lamprey. Cell Discov 3:17033
37. Carroll MC, Isenman DE (2012) Regulation of humoral immunity by complement. Immunity 37(2):199–207
38. Stephan AH, Barres BA, Stevens B (2012) The complement system: an unexpected role in synaptic pruning during development and disease. Annu Rev Neurosci 35:369–389
39. Kolev M, Kemper C (2017) Keeping it all going-complement meets metabolism. Front Immunol 8:1
40. Green DR (2017) Cell death and the immune system: getting to how and why. Immunol Rev 277(1):4–8
41. Green DR, Fitzgerald P (2016) Just so stories about the evolution of apoptosis. Curr Biol 26 (13):R620–R627

Further Reading

Cornejo E, Dchlaermann P, Mukherjee S (2017) How to rewire the host cell: a home improvement guide for intracellular bacteria. J Cell Biol 216:3931–3948
Green DR (2017) Cell death and the immune system: getting to how and why. Immunol Rev 277:4–8
Green DR, Fitzgerald P (2016) Just so stories about the evolution of apoptosis. Curr Biol 26:R620–R627
Janeway C (2017) Immunobiology, 9th edn
Leulier F, Lemaitre B (2008) Toll-like receptors – taking an evolutionary approach. Nat Rev Genet 9(3):165–178
Litman GW, Dishaw L (eds) (2013) Changing views of the evolution of immunity. Front Immunol 4
Loker ES (2012) Macroevolutionary immunology: a role for immunity in the diversification of animal life. Front Immunol 3:1–20
Ricklin D et al (2016) Complement component C3 – the "Swiss Army Knife" of innate immunity and host defense. Immunol Rev 274(1):33–58
Roers A, Hiller B, Hornung V (2016) Recognition of endogenous nucleic acids by the innate immune system. Immunity 44:739–754

Chapter 4
The Triumph of Individualism: Evolution of Somatically Generated Adaptive Immune Systems

The receptors of innate immune systems evolve slowly over time. Those that confer some fitness benefit will be naturally selected, and so become the common property of the succeeding generations. In contrast, the receptors of the so-called adaptive immune systems are generated somatically within each individual, by moving evolution from the level of the germline to that of somatic cells. As a result, each individual ends up with a repertoire of adaptive immune receptors that is as distinctive as are their fingerprints. Unlike the fingerprint, however, the repertoire of adaptive immunity in an individual is constantly changing. Adaptive immune systems come in two fundamentally different forms that differ both in the nature and in the source of the pathogen-sensing element. On the one hand are those immune systems that use nucleic acids as the pathogen sensors. In these cases the sensor is only formed after infection, and the key information needed to build it is derived from the pathogen. On the other hand are those "anticipatory" systems that use proteins as pathogen sensors. Here the sensors have been formed prior to infection with the pathogen, and the information used to form the sensors is entirely host derived.

Across the phylogenetic span from nematode worms to mammals, immune defence involves the use of a mix of "innate", i.e. germ line encoded, and of "adaptive", i.e. somatically encoded, receptors. Since different species inhabit different environments, it is clear that neither innate nor adaptive immunity can be viewed as an "off the peg" defence system. Though natural selection has ensured that certain features of innate and adaptive immunity are common between the fruit fly and mammals, there are also very considerable differences. In the same way, and for the same reasons, immune defence in the mouse is not identical to immune defence in humans.

© Springer Nature Switzerland AG 2019
R. Jack, L. Du Pasquier, *Evolutionary Concepts in Immunology*,
https://doi.org/10.1007/978-3-030-18667-8_4

4.1   Adaptive Systems that Use Nucleic Acid Sensors

The use of immune systems based on nucleic acid sensors preceded the division between germline and soma. Bacterial cells are germline and soma in one, yet they have evolved sophisticated adaptive systems of defence that go under the name of CRISPR-Cas.

4.1.1   Bacterial CRISPR-Cas

CRISPR-Cas systems provide bacteria with a means to protect themselves against viral infection. Since viruses evolve quickly, the selective pressure on the bacterial CRISPR-Cas systems has been enormous, so that now there are many different variants of the basic strategy. The essence of these systems is that a small part of a pathogenic viral genome is captured and inserted into the bacterial chromosome. The captured sequences can be transcribed and made available as "guide RNA" molecules that form a complex with a bacterial Cas nuclease. This complex of guide RNA and Cas protein functions as a weapon, in which the guide RNA identifies the complementary viral sequences, and the associated Cas nuclease then destroys the viral genome.

The CRISPR-Cas system is highly specific and enormously economic in terms of its requirement for space in the bacterial genome. The captured viral genomic fragment is small, and yet, once transcribed into a "guide RNA", it is large enough to act as a sequence-specific probe providing precisely tailored defence against a particular virus. However, there is a problem. Once a viral sequence has been captured, then the capturing bacterium—and all of its descendants—must clearly distinguish between the copy of the sequence in the attacking virus and the copy that now resides in its own genome. The sequence in the virus must be attacked; the sequence in its own genome must be left untouched. This is achieved by looking at the chromosomal sequences adjacent to the captured viral sequence. In some cases co-transcription of bacterial sequences that flank the viral insert results in a guide RNA containing both viral and bacterial sequences. Pairing of this guide RNA to the bacterial sequence aborts Cas-mediated destruction of the bacterial chromosome. The viral genome, on the other hand, lacks the "tolerising" bacterial sequence and is destroyed. Alternatively, in certain CRISPR-Cas systems, a "Protospacer Adjacent Motif", present in the viral genome, but not included in the sequence captured by the bacterium, allows the guide RNA-Cas complex to distinguish self from non-self.

4.1.2 Eukaryotic RNA Interference

The CRISPR-Cas systems are restricted to a large fraction of bacteria and archaea, but an alternative nucleic acid-based adaptive immune system is widely used in eukaryotes. This is based on the RNA interference machinery that is used in eukaryotes to help regulate gene expression. RNA interference involves small, so-called micro RNA (miRNA) genes, that are encoded in the genome, and which contain sequences complementary to short parts of the genes whose expression they will control. The transcripts are processed to yield short double-stranded miRNAs, and these are then incorporated into a large "RNA-Induced Silencing Complex" (RISC) in the cytoplasm. The miRNAs are analogous to the guide RNAs of the CRISPR systems in that they enable the RISC to specifically locate the target mRNA, which is then either destroyed or its translation is suppressed.

4.1.3 Evolution of Adaptive Immune Systems Based on RNA Interference

In the nematode worm *Caenorhabditis elegans*, this miRNA system has been adapted to provide defence against viruses. After infection of a cell, viral RNA is processed and loaded into a RISC complex that is then used to locate and destroy the virus. This adaptive defence system thus "borrows" genetic information from the viral pathogen and uses it to identify and destroy the intruder. Such a system works well within a single cell, but to be of real value in a multicellular animal, there has to be a little more. In particular, a way must be found to provide uninfected neighbouring cells with pre-emptive protection. In other words, this cell autonomous adaptive immune response must be converted into a systemic response. This achieved both in the nematode worm *Caenorhabditis elegans* [1] and in the fruit fly *Drosophila melanogaster* [2] by mechanisms that copy, and thus make large amounts of, the short viral RNA.

The formation of this "extra" RNA in the nematode and in the fly is one example of what is a recurring theme in the evolution of adaptive immunity. This is that two different species facing the same problem may reach similar solutions, but by quite different mechanisms.

Both the fly and the worm must convert a local RNA interference response into a systemic one, and in both cases this is achieved by amplification of the captured viral RNA and its export to other cells. However, in the case of the nematode, the amplification is by an RNA-dependent RNA polymerase, and the extra RNA produced in this way is then exported from the infected cell through special pores in the membrane (Fig. 4.1). Such pores, present on all of the cells, also permit entry of the RNA into cells at remote locations. The fly, in contrast, lacks both of these mechanisms, and instead amplifies the RNA using a reverse transcriptase borrowed from an endogenous retrovirus, after which the amplified product is released from

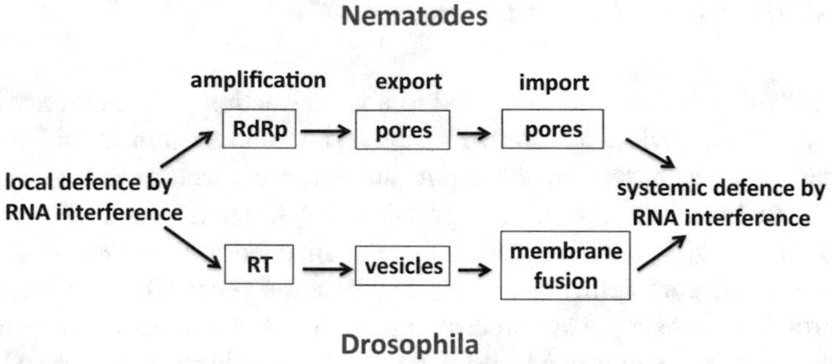

Nematodes

amplification   export   import

```
         ┌──────┐      ┌───────┐      ┌───────┐
    ──→  │ RdRp │ ──→  │ pores │ ──→  │ pores │ ──┐
         └──────┘      └───────┘      └───────┘   ↘
```

local defence by   systemic defence by
RNA interference   RNA interference

```
         ┌──────┐      ┌──────────┐      ┌──────────┐
    ──┐  │  RT  │ ──→  │ vesicles │ ──→  │ membrane │ ──→
         └──────┘      └──────────┘      │  fusion  │   ↗
                                         └──────────┘
```

Drosophila

Fig. 4.1 One problem—two solutions. Both in the worm and in the fly a local RNA interference response must be converted into a systemic one. In the nematode (upper line) the RNA is amplified by an RNA-dependent RNA polymerase (RdRp), exported from the infected cell through pores, and imported into distant cells by pores. In the fly (lower line) amplification is by a reverse transcriptase (RT), the product is packed and exported from the infected cell in vesicles, which deliver the RNA to distant cells by membrane fusion

the infected cells in vesicles, which can deliver it to uninfected cells at remote locations (Fig. 4.1). In both cases the RNA interference response in infected cells provides distant uninfected cells with pre-emptive protection against the virus.

At least in the case of nematode worms, this adaptive response is one of those situations in which the accepted norms of evolutionary thought have to give way to the pressing necessities of immune defence. One widely held view is that the strict separation of germline and soma (Sect. 2.1) makes the Lamarckian idea of the inheritance of acquired characteristics impossible. However, in *Caenorhabditis elegans* the oocytes, like the somatic cells, are able to take up the exported RNA and, as a result, the worm's generation of immunity to a virus can be transmitted as an acquired characteristic to its progeny [1].

In vertebrates endogenous miRNAs play important roles in controlling gene expression, but this sort of RISC-based system does not play a significant role in defence against viruses. This is something of a puzzle, for the adaptive defence systems in worms and insects based on RNA interference are very ancient, and evolution does not readily give up a good idea. The currently most widely accepted thought is that a good idea becomes dispensable if a better one comes along. The better idea seems to have been to replace the "adaptive" RNA interference response of invertebrates with the "non-adaptive" type 1 interferon response, which is used as an antiviral defence mechanism in the innate systems of fish, birds and mammals [3]. These type 1 interferons are cytokines that are induced as a result of the activation of many of the receptors of innate immunity. They cause changes in the expression of hundreds of genes, many of which have potent antiviral activity. The simultaneous induction of multiple antiviral functions makes it difficult for a viral pathogen to readily evolve escape mechanisms [4]. In mammals, the interferon system has been backed up with the emergence of the APOBEC3G cytidine deaminase, which is involved in providing innate defence specifically against retroviruses (see Sect. 3.6.1), and by the evolution of T-cell-based adaptive immunity (see Sect. 4.6).

4.2 Somatic Evolution of Immune Systems that Use Protein Sensors

Perhaps the simplest way to form a protein-based adaptive immune system would be to make use of the existing array of innate receptors and then, when under pathogen attack, to randomly mutate their genes in the hope of making a somewhat better receptor. Something along these lines may take place in the snail *Biomphalaria glabrata*. When attacked by parasitic worms the snail's FREP receptor genes seem to pick up random somatic mutations, which may aid in fending off the parasite [5]. This way of doing things, however, was certainly never more than a footnote in the history of immunity, and effective protein-based adaptive immune systems had to await the evolution of vertebrates.

Vertebrate anticipatory adaptive immune systems based on protein receptors are radically different from the adaptive RNA interference system of invertebrates. The crucial difference is that in the RNA-based systems the information that provides the receptor with its specificity is "borrowed" from the pathogen, while in protein-based systems this information is derived solely from the host. RNA-based adaptive immunity, and the innate systems looked at in Chap. 3, can be fairly thought of as straightforward evolutionary answers to the challenge of detecting and destroying pathogens. Natural selection established these defence systems, over the course of hundreds of millions of years, by screening many billions of random mutations. At each generation the few advantageous changes were favoured, while those individuals expressing mutations that would lead to disaster were selected out. Protein-based adaptive immune systems, in contrast, face the problem that they must carry out this process of mutation and screening of the repertoire "all at once". The somatically generated repertoire is made in a hurry and will contain not only useful receptors, but also many worthless or even autoimmune ones. In fact this repertoire will certainly be lethally autoimmune unless some way can be found to purge it of these deleterious specificities. Purging the repertoire by somatic "tolerance" mechanisms is therefore crucial to the establishment of an anticipatory adaptive immune system. Only after all this has been done can the repertoire be used to deal with pathogens.

The difficulty for a protein-based immune system is that in order to be able to detect all possible pathogens, it must produce a truly vast number of different receptors. How can such a huge receptor repertoire be formed in the germline on a genome that has space for only around 20,000 genes? The simple answer is that it can't be done—at least not directly. However, evolution has found a trick to solve this problem. The trick is that the antigen-binding receptor genes are not encoded as such in the germline, but instead are assembled in the developing lymphocytes by mixing and matching sets of pre-existing sequence modules at the DNA level. Useful receptors, made in this way, cannot be passed on to the next generation, because the receptor genes are not formed in cells of the germline. Instead the next generation inherits the collection of DNA modules, and the means to join them together, as a "do-it-yourself kit" with which it too can somatically build receptor genes.

This sort of deliberate induction of mutations in somatic cells is normally avoided, because it brings with it the risk of cell death—or even worse—of cell transformation. Despite this, lymphocytes make use of a potent, induced mutagenic process to form the antigen-binding domain of their immune receptors. Indeed the mutational process induced in these cells is so effective that every new lymphocyte that is generated carries a mutant receptor, and every one of these mutant receptors has a unique binding site. One great advantage of this way of doing things is that, since the mutation process directly alters the DNA sequences coding for the receptor's binding site, only a small investment of genomic space is sufficient to allow for the generation of many different receptors.

There is, it must be said, one aspect of this story, which goes against our usual expectations of how evolution works. One tends to think of evolution as being slow and conservative—a matter of finding some sort of solution to a current problem, and then "tinkering" with that solution till it is gradually improved. That is probably a reasonable picture of how things are done—most of the time. What is most unusual is for evolution to suddenly change course, or for a solution that works to be simply dropped, and replaced with something completely different. Two such exceptional events can be seen in the evolution of the antigen-specific receptors of adaptive immunity. In the first case a system of somatic receptor construction "suddenly" arose in early jawless vertebrates (agnathans) and gave rise to a huge repertoire of receptors composed of Leucine Rich Repeat (LRR) domains. The second abrupt change of direction occurred when this agnathan system was subjected to massive re-organisation and emerged as something rather different in the jawed vertebrates (gnathostomes), where the receptors are all composed of "Immunoglobulin Super-family" (IgSF) domains (see Appendix D).

4.2.1   Two Different Protein-Based Adaptive Immune Systems

There are two systems of protein-based adaptive immunity in vertebrates. The first is found in the agnathans, whose surviving members are the lampreys and hagfish. The second is found in the gnathostomes, which comprise the vast majority of all vertebrate species. In this chapter we will look at the evolution of function in these two adaptive systems. One must, however, bear in mind that while adaptive immunity in gnathostomes has been studied for well over a century, the analysis of the nature of the agnathan adaptive immune system receptors only began with a seminal publication in 2004 from Max Cooper's laboratory [6]. Not surprisingly, while there is a mass of detailed knowledge on gnathostome adaptive immunity, very little, by comparison, is known about the workings of the system in agnathans.

4.2.2 Lymphocytes and Their Receptors

The lymphocytes that somatically form immune receptors in vertebrates can be divided into two broad types. The first consists of those that, once activated by contact with a pathogen, will secrete their receptors in the form of highly specific "antibodies". These are the familiar B-cells of jawed vertebrates and the VLRB cells of agnathans. The second type of lymphocyte—best known here are the $\alpha\beta$ T-cells of jawed vertebrates and the VLRA and VLRC cells of agnathans—is characterised by the fact that they do not secrete their antigen-specific receptor. Since these two different lymphocyte forms have different functions, it is not surprising that different selective forces act on the formation of their receptor repertoires. However, some general aspects of receptor structure and generation are common to them all.

Any antigen receptor expressed on the surface of a lymphocyte has to be structurally robust, so that it can survive in the frequently difficult extracellular environment. At the same time, the ligand-binding part of the structure must be able to accommodate many variations in its sequence so that many different versions of the receptor can be formed. In reality these criteria are not terribly restrictive, and a number of different protein domains would fit the bill. The LRR domain used in agnathan immune receptors, and the IgSF domain used in gnathostomes are found throughout animal phylogeny, and both frequently occur in proteins expressed on cell surfaces. Irrespective of whether the receptor is of the LRR or of the IgSF type, the ligand-binding sites of all adaptive immune receptors in vertebrates are formed somatically during lymphocyte differentiation, by mixing and matching DNA elements from a library of sequence modules in the genome. The advantage of this way of doing things is clear: if the sequence coding for the antigen-binding site of a receptor is built up by bringing together elements from a collection of short sequence modules, then the number of different receptors that can be made greatly exceeds the number of modules. If this were just a numbers game, then there would be no difficulty at all in arranging for the construction of thousands of billions of different receptor-binding sites, for all that would be required would be lots of alternative module sequences. 10^4 alternative modules would do the job easily—and this would not take up an inordinate amount of space in the genome. However, each of these modules would have to be continually subject to purifying natural selection, otherwise random mutation would quickly destroy their coding capacity. Sadly, the simple truth is that the larger a family of functionally similar sequences becomes, the more difficult it is for natural selection to maintain all of the members intact. With the available selection pressure spread over an increasing number of sequences, random mutations leading to stop codons, frame shifts or disruptive amino acid replacements will inevitably accumulate, and this sets strict limits on the number of modules that can be maintained.

So now we are back to square one. If the number of useful modules is restricted in this way, how can a large number of receptor sequences be generated? The answer is that the problem only arises if the module-joining process is precise, so that joining module 1 to module 2 always yields exactly the same result. If, however, the recombination process were to be sloppy, then joining module 1 to module 2 may yield any one of many different results. In both of the protein-based adaptive systems that we know of, the mechanism to mix and match modules is extremely sloppy, and because of this, the repertoire of different antigen-receptor-binding sites that can be formed is enormous.

4.3   Somatic Formation of the Agnathan Adaptive Immune Receptors

Agnathan immune receptors—the so-called variable lymphocyte receptors (VLRs)—are constructed out of LRR domains. In most cells the genes for these receptors are massively incomplete and consist just of the sequences coding for the beginning and end of the polypeptide chain. The central parts of the gene are only formed in developing lymphocytes, which copy in LRR modules from a collection of flanking sequences (Fig. 4.2). Simply copying in members of the flanking collection of LRR modules would not be enough, for the required large receptor

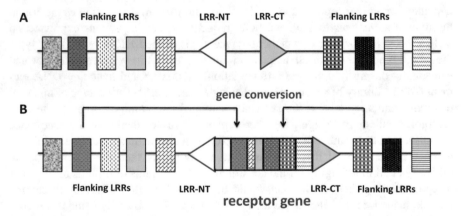

Fig. 4.2 Agnathan adaptive immune receptor gene rearrangement. (**a**) Germline configuration of a lamprey variable lymphocyte receptor gene. The gene codes only for the beginning—N-terminal LRR (LRR-NT)—and the end—C-terminal LRR (LRR-CT) of the receptor (shaded triangles). These are separated by a "spacer" sequence. This almost "empty" gene is flanked on each side by many sequences coding for leucine rich repeat (LRR) domains. (**b**) Flanking LRR sequences are copied by a gene conversion mechanism into the empty gene where they replace the spacer sequence. The gene conversion process is believed to be initiated by a cytidine deaminase enzyme similar to the mammalian "Activation-Induced Deaminase" (AID). The copying mechanism permits the formation of "hybrid" LRR domains copies. The number of flanking sequences remains unaltered

repertoire can only be made if the copying process is imprecise. In agnathans the process is almost endlessly imprecise, because the flanking LRR sequences are not always treated as individual elements. Instead, copying can start in one module and then, somewhere in the middle, jump to another one. In this way, an essentially unlimited number of "hybrid" LRR sequences can be formed from a restricted library of flanking LRR sequences. In agnathans each lymphocyte is limited to expressing one receptor specificity. It is not known how this is imposed, but an analogous phenomenon exists also in gnathostome adaptive immunity (see Sect. 4.5.4).

4.3.1   The Key Enzyme: An AID-Like Cytidine Deaminase

Central to the formation of the lamprey adaptive immune system receptor repertoire, is the ability to somatically construct hybrid LRR receptors from a limited set of germline encoded LRR modules. There are three loci containing genes for the receptors in the agnathan genome. Each individual lymphocyte uses only one of these genes, and so three different types of lymphocyte are formed. These lympho-cytes express either the "Variable Lymphocyte Receptor A" (VLRA), or VLRB or VLRC.

Any mechanism to mix and match DNA modules requires the ability to cut DNA. In the agnathan system, the cutting process is initiated by an enzyme with cytidine deaminase activity. Cytidine deaminases of the AID/APOBEC family (Sect. 3.6.1) arose early in vertebrate evolution. Those members of the family that are involved in adaptive immunity remove the amino group from dC residues in the single-stranded stretches of DNA, which may be formed as a result of the supercoiling stress induced during transcription that locally "melts" the double helix. The deamination converts dC to dU, a base that is not accepted as a normal component of DNA by the cell's DNA repair mechanisms [7]. The repair systems recognise these dU bases as "errors", and set about removing them. There are a number of different repair systems available, and the result of the repair of the "error" depends on which of them operates. Little is known as yet about the mechanisms that regulate the formation and the repair of dU in DNA in agnathans, but the result of their removal is a single-stranded break in the DNA—and this is the first essential step in the gene conversion process.

4.3.2   Structure of the Receptor Molecules

VLR molecules, like all LRR proteins, form horseshoe-shaped structures similar to those formed by TLRs (Sect. 3.3), though the VLR molecules have fewer LRR domains and consequently form a shallower crescent-shaped structure (Fig. 4.3).

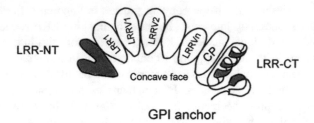

Fig. 4.3 Diagrammatic outline of a VLRB molecule. The LRR domains (LRR) and the connecting peptide (CP) (open ovals) form a curved structure, which is bounded at the N-terminal end by the "capping" domain LRR-NT, and at the C-terminal end by LRR-CT. Antigen is bound on the concave face and the LRR-CT also plays an important role in this interaction. The VLRB molecule is attached to the cell surface by a glycophosphatidylinositol (GPI) anchor [8]

Strikingly, they bind antigens in the inner, concave, face of the molecule, rather than on the outer face as do the TLRs. This type of structure provides a large area of interaction with a ligand, and results in an affinity that is comparable to that between a gnathostome antibody and its antigen.

The functions of the VLRA and VLRC bearing cells are at the moment unclear. However, the VLRB expressing cells, once activated, secrete their receptor, and since the secreted form of the VLRB molecule is a complex of eight or ten such units, the effective affinity (avidity) of the complex is substantial and may rival that of a mammalian pentameric IgM molecule.

4.3.3   Tolerance

"There's no such thing as a free lunch", said Milton Friedman, and this slogan is as true in immunology as it is in economics. A vast receptor repertoire, which "sees" everything, is splendid—in terms of defence against pathogens, but on the other hand such a repertoire "sees" far too much, for it will undoubtedly detect all the myriad components of the organism's own body. The simple—and correct—conclusion is that any usefully large repertoire, constructed by a process of random mutation, is bound to be lethally autoimmune. An adaptive immune system therefore has to co-evolve along with some powerful mechanisms that remove or suppress the receptors with autoimmune specificity. Though such "tolerance mechanisms" must of necessity exist in the agnathans, nothing is as yet known about how they work.

4.4 Adaptive Immune Receptors in Gnathostomes

In jawed vertebrates lymphocytes also express antigen-specific receptors and these are also somatically formed by DNA rearrangement during lymphocyte development. This sounds like a rehash of the agnathan story—but it is not, for in jawed vertebrates the antigen-specific receptors are composed of a different protein domain, and a different mechanism is used to mix and match the DNA modules that code for them.

Like agnathans, gnathostomes also have three major categories of lymphocytes expressing adaptive system receptors. The first are the "B-cells" which express an antigen-specific B-Cell Receptor (BCR), and which, once activated, may be induced to secrete the receptor in the form of soluble antibody. These B-cells of gnathostomes are thus reminiscent of the VLRB cells in agnathans. The second major category consists of the "CD4$^+$ T-cells", and the third of the "CD8$^+$ T-cells". These latter two cell types express a somatically generated antigen-specific T-cell receptor (TCR) that is not released in soluble form. These T-cells are, in this sense, reminiscent of the VLRA and VLRC cells of agnathans. All of these antigen-binding receptors of gnathostomes are composed of "IgSF" domains (Fig. 4.4; Appendix D).

4.4.1 The Immunoglobulin Super Family Domain

Many proteins contain domains that are members of the IgSF. These domains fall into four structural sets: the variable set (V), the intermediate set (I) and the constant sets (C1 and C2). V, C2 and I type domains are found in practically all metazoans, while C1 domains are found only in gnathostomes (see Appendix D). In the antigen-specific receptors of gnathostome B-cells and T-cells, the N-terminal V-type domain is always associated with one or more membrane-proximal C1-type domains. The antigen-binding part of these receptors are all derived from one particular V-type domain, which emerged as a crippled casualty from an ancient battle fought out some 500 million years ago between the precursor of all jawed vertebrates and a DNA transposon.

4.4.2 The "Transib" Transposon Contributed to the Structure of All Antigen-Specific Receptors in Jawed Vertebrates

Transposons are short stretches of parasitic DNA that consist, at a minimum, of terminal sequences that flank a gene coding for a "transposase". The transposase is a site-specific recombinase that recognises the terminal sequences and can catalyse the precise excision of the transposon out of a genome, and its insertion into a new position. Since a transposon can use this "cut and paste" mechanism to hop around,

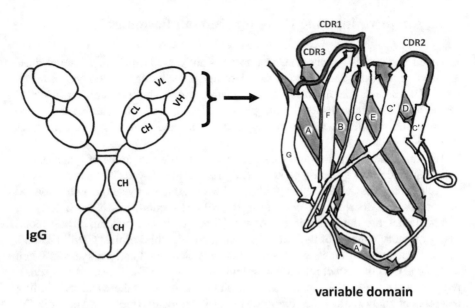

variable domain

Fig. 4.4 Immunoglobulin super family domains in an antibody. Left: schematic diagram of a gnathostome IgG molecule. The molecule is composed of four chains, two identical heavy (H) chains and two identical light (L) chains. Each H consists of one N-terminal "variable" (V-type) domain (VH) followed by "constant" (C1-type) domains. The L chain consists of one V-type domain (VL) followed by one C1-type domain. The antigen-binding part of the molecule is formed by the association of VL and VH. Right: a ribbon diagram of the structure of a V-domain. The domain is formed of two sheets of β-strands and is reinforced by a disulphide bond formed between strands B and F. Loops linking strands B and C, C' and C'', and F and G form the "Complementarity Determining Regions" (CDR1, CDR2 and CDR3) that make contact with the antigen (see also Fig. 4.7)

there is always a risk that sooner or later it will hop into, and disrupt, an essential gene. Because of this, there is considerable selective pressure to make sure that transposons are quickly immobilised, either by inactivation of the transposase or by destruction of the terminal repeats. There are currently no active transposons in humans, though around 3% of the genome is composed of their inactivated remains. One such inactivated transposon belonged to a member of the Transib (Trans Siberian) transposon family [9].

The adaptive immune receptors of jawed vertebrates were born when a Transib transposon, (Fig. 4.5) [11], inserted into an exon coding for a V-type IgSF domain. In the aftermath, the recombinase genes and the terminal repeats from the transposon were hijacked, and retooled to produce the lymphocyte receptors. What seems to have happened is that the coding sequences for the transposase genes were removed, and placed elsewhere in the genome, where they sit under the tight control of a lymphocyte-specific promoter. These genes are now known as RAG-1 and RAG-2. The terminal inverted repeats, however, were left in place within the V-domain. In

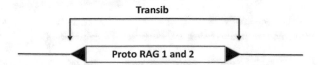

Fig. 4.5 The Transib transposon. A Transib transposon integrated into a chromosome. The transposon consists of genes coding for the transposase genes ProtoRAG-1 and ProtoRAG-2 flanked by terminal inverted repeats [10]. The ProtoRAG recombinase excises the transposon precisely by cutting the DNA at the ends of the terminal repeats

such a situation expression of the recombinase genes will cause precise excision of the terminal inverted repeats and of the sequences lying between them.

However, it must be admitted that a very great deal has happened in the last 500 million years, and the chromosomal segments, which contain the genes for adaptive immune receptors, are the most complex, segmentally duplicated regions in the human genome [12]. Many steps of duplication and insertion, involving the terminal inverted repeats, were required to form the locus, and at every step natural selection would have determined whether the changes were worth preserving. It will be clear that going from the initial insertion of the transposon to a functional receptor locus would have required a considerable period of time. It is unclear how adaptive immunity was organised over this transitional period (see also Sect. 4.16.7).

The two major lineages of lymphocytes of adaptive immunity in jawed vertebrates—the B-cells and the T-cells—express quite different receptor structures, which are coded for at different loci in the genome, but all of the antigen-binding sites are derived from the original V-type domain, into whose gene the Transib transposon inserted.

The process by which the gene segments coding for these V-domains are formed during lymphocyte development is radically different from that used to build the agnathan adaptive immune receptors. In the lamprey, information is copied by gene conversion from the flanking sequences into the empty receptor gene. In gnathostomes the information is physically moved from one position to another by a recombination process catalysed by the RAG genes (Fig. 4.6). However, just as in the agnathan system, the mere recombination of a limited set of modules is not sufficient to produce a huge repertoire of different receptors and in the gnathostomes the recombination process has been modulated so as to introduce the necessary degree of sloppiness into the joining of the V, (D), and J modules.

4.4.3   The Gnathostome Immune Receptor Antigen-Binding Sites

The V-type IgSF protein domain, encoded by the gene that was invaded by the Transib transposon, consists of a stable central structure from which extend three loops (Fig. 4.4) [13]. As outlined in Fig. 4.7 these loops form the site at which the

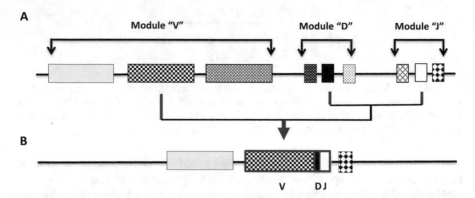

Fig. 4.6 Generation of gnathostome adaptive immune receptors from V, D and J modules. (**a**) Germline configuration. The example shown illustrates in schematic form the situation in the heavy chain gene of the BCR and in the β chain of the TCR. The germline gene contains separate V, D and J modules each of which is flanked by Transib-derived terminal repeats. Genes coding for the BCR light chain or for the α chain of the TCR are similar but lack the D module. (**b**) The rearranged gene in this example is formed by RAG-mediated recombination of one D with one J module, followed by recombination with one V module

receptor physically interacts with the accessible surface of an antigen. Since the loops, or parts of them, must form a structure complementary to the surface of the antigen to which they will bind, they are known as the "Complementarity Determining Regions" (CDRs). With rare exceptions, the antigen-binding sites of gnathostome adaptive immune receptors are constructed from two separate V-type IgSF domains. The resulting heterodimer forms a binding site that bears a total of six CDRs (Fig. 4.7). The use of heterodimers to form the binding site substantially increases the repertoire size, for 10 monomers of each chain can, in principle, form 100 different receptor specificities.

4.5   RAG and Its Limitations

A glance at the chapter on V, (D), J recombination in any immunology textbook will quickly convince you that many proteins other than the RAG recombinases are involved in a process that is seemingly bizarrely complex. Why should this be so? The answer is that precise joining of the handful of V, D and J modules available would yield—at best—a modest repertoire of receptors. A protein-based immune system, however, is of little or no selective value, unless the receptor repertoire produced is vast. RAG, which we "inherited" from the Transib transposon, is a site-specific recombinase, and so the complicated molecular gymnastics, described in the RAG recombination chapter of your immunology textbook, are there for the simple purpose of converting a precise site-specific recombination process into a sloppy one. The result is very sloppy indeed: DNA is cut, ends are nibbled, single-stranded

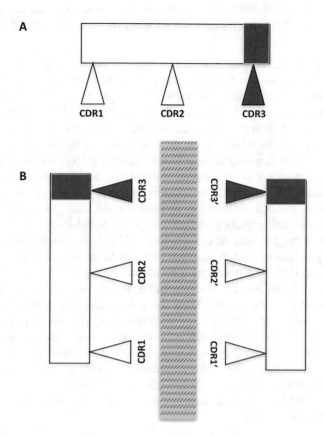

Fig. 4.7 The antigen-binding V-type Immunoglobulin Super Family (IgSF) domain of gnathostome adaptive immune receptors: (**a**) The V-type IgSF domain has a stable backbone of conserved sequence that supports three variable sequence loops. These form the three Complementarity Determining Regions (CDRs), which make contact with the antigen (stippled rectangle). The parts of the receptor derived from the "V" module only (Fig. 4.6) are shown in white (CDR1 and CDR2). The part of the molecule derived from the rearranged sections of the "V", ("D") and "J" modules is shaded (CDR3). (**b**) With few exceptions, the antigen-binding sites of gnathostome adaptive immune receptors are composed of two different chains. In the BCR these are the heavy (H) and light (L) chains, while in the TCR they are the α and β chains. Thus both BCR and TCR have six different CDRs involved in ligand binding

overhangs are filled in or chopped off, bases are moved from one strand to the other, or are even randomly added on to the broken DNA ends. It is a mess, but it is a mess that allows for a very large repertoire of products to be formed.

4.5.1 Somewhere Between Lots of Receptors and Lots of Junk

The sloppy RAG recombination system is there to make lots of different receptor structures. This it does, but it is important to keep in mind that the RAG recombination system is quite astonishingly wasteful. A large part of the problem is that the genetic code is a triplet code, meaning that three bases in DNA sequence will be translated into one amino acid in the sequence of a protein chain. Yet RAG recombination, with its nibbled ends, filled in gaps and its anarchistic addition of random nucleotides, is unable to count bases. Whether multiples of one, two or three bases have been added in, or chopped out, during the recombination process, is simply a matter of chance. Yet all adaptive receptor protein chains consist of a V-type domain, which is followed by at least one essential C-1-type IgSF domain (Fig. 4.4). If the reading frame is shifted during RAG-mediated recombination, then the structure of the downstream C-1 domain will be destroyed. On this score alone 66% of the products of each RAG recombination event are going to end up being molecular junk. But that is not the end, for even if the reading frame is maintained a chaotic mutational process may generate stop codons, or sequences coding for amino acids that disrupt the three-dimensional structure of the protein chain. The bottom line is that something like 70% of each of the recombination events required to form a gnathostome V-type domain will end in catastrophe. Furthermore, with few exceptions these gnathostome adaptive receptors are heterodimers. Thus the 30% of lymphocytes emerging from RAG recombination with the first chain intact must now rearrange the second chain. Once again 70% of these rearrangements of the second chain will result in junk. This is a degree of waste, which natural selection must have viewed critically.

4.5.2 RAG Recombination: A Half-Hearted Attempt to Build an All-Encompassing Adaptive Repertoire

RAG recombination does make lots of different receptors by messing about with the DNA coding for CDR3, but it leaves CDR1 and CDR2 untouched (Fig. 4.7). Why this hesitation? The answer is that if CDR1 and CDR2 were also to be mutated in the same way, then all of the problems associated with rearrangement at CDR3 would be multiplied by three, and essentially all of the rearranged receptor genes would be junk. This must be avoided at all costs, and the price, which must be paid, is the reduction of the completeness of the RAG-generated repertoire.

4.5.3   Ligand Binding and "Tolerance" in the B-Cell Lineage

The "sloppiness" of the rearrangement process, which is the secret of adaptive immunity's success, both in agnathans and in gnathostomes, automatically forms a receptor repertoire that will contain numerous autoimmune specificities. These must be removed, one way or another, by "tolerance" mechanisms—and the jawed vertebrates' concern with tolerance can be fairly described as obsessive. Not surprisingly, the study of tolerance mechanisms currently forms the forefront of immunological research, because understanding how tolerance works will provide the keys to manipulate immunity in situations as varied as vaccination, autoimmunity, cancer and the treatment of infectious diseases.

Given the random generation process, the BCR receptor repertoire can "see" essentially any and every molecular structure. In particular it can "see" all of the structures in our own body. Experimental analysis of the specificities of newly formed BCRs suggests that approximately 50% of them are directed against "self" structures [14], and that many of these receptors are removed by a "central tolerance" system, operating on the newly formed B-cells in the bone marrow and spleen [15]. Central tolerance deletes the autoimmune specificities by killing the newly formed cells that express them.

The requirement for central tolerance has enormous consequences for the way adaptive immune receptors are expressed. Innate cells like macrophages express many different types of innate system receptors. In contrast, each B-cell expresses a single type of receptor molecule, which consists of two copies of one "heavy" (H) chain and two copies of one "light" (L) chain (Fig. 4.4). Suppose that a B-cell were to express two different heavy chains (H1 and H2) and two different light chains (L1 and L2), then it could express four different combinations (H1L1, H2L2, H1L2 and H2L1), and hence four different receptor specificities. You might think that this would be a good idea, for such a B-cell could now perhaps recognise four different pathogens, but in reality it would be terrible. The reason is very simple: if 50% of B-cells that express one single receptor must be destroyed by central tolerance because they are autoimmune, then 75% of any, which expressed 2 different receptors, 88% of those expressing three receptor specificities and 94% of those expressing four specificities would have to be destroyed. By permitting each lymphocyte to express only one receptor, the system maximises the number of B-cells that will survive the central tolerance processes.

4.5.4   Allelic Exclusion

Limiting each lymphocyte to just one receptor is achieved by the unusual genetic process of "allelic exclusion". In eukaryotes, there are two copies of each autosomal gene—one maternal and the other paternal. In general, both alleles are expressed at an approximately equal level, but there are a few situations in which pressing selective forces drive the preferential use of either the maternal or the paternal copy. This sort of "allelic exclusion" phenomenon is seen, for example, in mice and humans in the expression of imprinted genes involved in resolving "parental conflict" [16]. It also occurs, though by a different mechanism, in the expression of odour receptor genes in the nasal neuroepithelium, which require monoallelic expression to achieve correct wiring with the CNS [17]. The recombination of lymphocyte receptor genes is a third example. RAG rearrangement is completed first on just one of the two alleles and, if this is successful, then rearrangement on the second allele is blocked. If, however, the rearrangement on the first allele fails to produce a well-formed receptor chain, then the cell is given a second chance by being allowed to try again on the second allele. The result is that, in general, each B lymphocyte expresses only one single receptor specificity.

4.6   Ligand-Binding in the T-Cell Lineages

The major T-cell lineages express heterodimeric "$\alpha\beta$" TCRs. These $\alpha\beta$ T-cells differentiate from precursors that are born in the bone marrow, and then migrate to the thymus where the RAG-dependent somatic generation of the TCR takes place. Gene segments, homologous to the V, (D) and J modules of the BCR, encode the ligand-binding domains of the heterodimeric TCR. The overall structure of the TCR's antigen-binding domains is similar to that of the BCR. Each of the antigen-binding domains contains three CDR loops that make contact with the ligand (Fig. 4.7). Despite the fact that only the CDR3 loop is mutated during RAG recombination, a huge repertoire of different TCRs is generated. As with the BCR, the rearrangement of the TCR is subject to allelic exclusion so that each T-cell normally expresses only one receptor specificity.

These TCRs are quite different from the receptors of B-cells in terms of the sorts of ligands that they can bind. A BCR may be complementary to a small molecule, or it may recognise part of a large molecule such as a protein, or polysaccharide, or it may even recognise part of a multi-molecular complex, such as a virus, a bacterium or a eukaryotic cell. In contrast, the vast majority of $\alpha\beta$ TCRs recognise short peptides. These peptides have to be first formed inside some other cell, then complexed with peptide-binding proteins, before being displayed on the cell's surface. What is the point of this? At least part of the answer has to do with the fact that some bacterial, and all viral pathogens, live and replicate inside cells where the antibodies produced by B-cells cannot reach them. One principle function of T-cells is to combat such intracellular infections—but therein lies a major problem. The T-cell's antigen-specific receptor is bound on its surface membrane, pointing out into the extracellular space. If these receptors are to be of use in detecting and combatting pathogens lurking inside other cells, then they must be able to monitor and unobtrusively "see" inside each of the approximately 10^{14} cells in an adult human being. How can this be done?

This problem is similar to that faced by the world's intelligence agencies, all of whom wish to know what each of us is doing. They cannot regularly interrogate all of the roughly 8×10^9 human beings on the planet, and so, as Edward Snowden showed, they use unobtrusive means to come by the information they require. Mobile phones, laptops, credit cards and car navigation systems are the commonplaces of modern life, and each time one of these devices is used, the information is swept up by the security services, and subjected to sophisticated data analysis to decide what, if anything, needs to be done to whom. This is the strategy used by T-cells. In place of the data from electronic devices, they make use of information extracted from the pattern of proteins that are present within a cell. Know that pattern, and you know, with considerable precision, the intimate details of what is going on inside that cell. In a nutshell, this is done by digesting a small fraction of the proteins within a cell into short peptides, which are bound by peptide carrier proteins, and the resulting peptide–peptide carrier complexes are then expressed on the cell surface, where they may be "seen" by circulating T-cells.

4.6.1   The MHC Complex and the Peptide Carrier Molecules

Winston Churchill's phrase "a riddle, wrapped in a mystery, inside an enigma" was not directed at the Major Histocompatibility (MHC) complex, but it does provide a fair description of it. Two rounds of whole genome duplication early in the evolution of vertebrates have left four identifiable copies of this region in humans with one copy normally present on each of chromosomes 1, 6, 9 and 19. The bona fide human MHC complex, containing the genes for the peptide-binding proteins, is the copy on chromosome 6 [18]. Long before the evolution of adaptive immunity, this ancient section of the genome contained a number of genes involved in innate immunity, stress responses and protein degradation [19]. The riddle of the MHC concerns this "primordial immune complex" [20], for though there are variations in gene content, relative position and chromosomal location throughout phylogeny, many of the genes within it have indeed been kept together for a very long time. Is there a selective advantage in this maintenance of gene linkage? One explanation has been that it was all a matter of chance, for once genes have arisen on a chromosome they will stick together until some chromosomal rearrangement like an inversion or translocation separates them. Since such events are rare, genes and gene order (synteny) can often be traced across vast tracts of phylogeny. Alternatively, the genes may have been kept together for some functional reason. For example, since different chromosomal domains have different degrees of accessibility for the transcriptional apparatus, keeping genes that code for related processes together may be of selective value if it helps ensure their concerted expression [21]. One further possible mechanism is based on the interactions of polymorphic genes. Imagine two genes, "A" and "B", whose protein products must interact. Let's assume that both of these genes are polymorphic, so that different individuals in the species have different versions of "A" and different versions of "B". If for each of

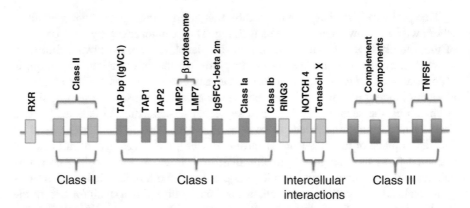

Fig. 4.8 A hypothetical scheme of the "Ur-MHC" of gnathostomes. The genes can be grouped into four categories: the Class-I region, which includes the Class-Ia and Class-Ib genes together with the LMP genes coding for proteasome subunits, and the TAP (transporter associated with peptide antigen processing) genes involved in the generation and transport of the peptide ligands for Class-I (red elements); the Class-II region containing the genes for the Class-II molecules (orange elements); the Class-III region that contains a number of genes involved in innate immunity (blue elements), such as the complement components C2, C4 and B, and cytokines of the tumour necrosis superfamily (TNFS). In addition, other genes including NOTCH and Tenascin X (green elements) are encoded in the MHC complex. Little of this hypothetical structure now remains unchanged across all gnathostome groups. Class-I and Class-II genes are not necessarily linked, β2 microglobulin is present in the MHC of modern sharks, but not elsewhere, and the Class-II genes themselves are, under certain circumstances, dispensable, as in the Atlantic cod

the polymorphic "A" genes there is an ideal partner "B", then there will be a clear advantage in keeping each ideal "A + B" pair tightly linked in the chromosome [22].

Beyond this riddle of the "primordial immune complex" lies the mystery surrounding the forces that, early in jawed vertebrate evolution, drove the selection of the genes coding for the peptide-binding MHC proteins. A hypothetical structure for the resulting early gnathostome "Ur-MHC" complex is shown in Fig. 4.8. Numerous speculative scenarios have been developed, but as yet there is no consensus as to where these genes came from, and how their evolution was matched to that of the TCR. Not surprisingly, given these imponderables it remains an enigma whether the presence of the genes coding for these peptide-binding molecules in this particular locus is evolutionarily significant or merely a case of "they had to go somewhere".

Though little is known about the early evolution of the peptide-binding MHC molecules, a great deal is known about how they function in vertebrate adaptive immunity. These molecules come in two forms—Class-I and Class-II—which differ not only in structure (Fig. 4.9), but also in the nature of the peptides that they bind. Class-I molecules bind peptides, which are usually nine amino acids long. Class-II molecules bind longer peptides, but even here the binding energy is derived largely from the nine amino acids that fit into the binding groove. The number of possible peptide ligands for these MHC molecules is large. Since there are 20 amino acids, the number of different nonamer peptides that could, in principle, be produced is 20^9,

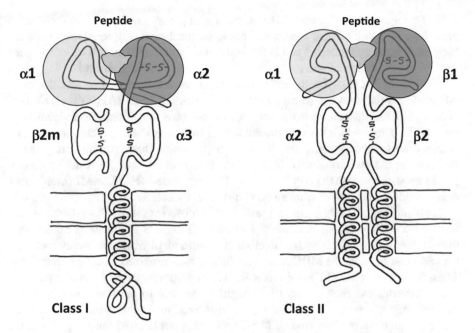

Fig. 4.9 Class-I and Class-II MHC molecules. Class-I MHC molecules (left) consist of three domains: the α1 and α2 domains, which together form the peptide-binding groove, and the C1 type IgSF domain α3. The α3 domain interacts non-covalently with beta-2 microglobulin, and this interaction is important for the stability of the complex. The molecule is bound to the cell membrane through a transmembrane domain. The Class-II MHC molecule (right) is composed of two chains—α and β—each of which is built of two domains. The α1 and β1 domains together form the peptide-binding groove. The Class-I and Class-II molecules have remarkable structural similarities. The peptide-binding grooves of both are similar, and both have an IgSF domain of the C1 type that associates with the membrane. The homology between the two classes is indicated by similarity of shape and shading. The Class-I α1 domain is homologous to Class-II α1. The Class-I α2 domain is homologous to Class-II β1. Class-I α3 is homologous to Class-II α2 and β2 as well as to β2 microglobulin

which is roughly 5×10^{11}. Though not all of these peptides will actually be formed, because the proteases responsible for making them do have preferences as to where to cut a protein chain, yet even so the array of peptide available to bind to MHC molecules, is large.

4.6.2   Why Two Types of MHC-Peptide Carrier Molecules?

Every nucleated cell in the body digests a fraction of its proteins in a cell organelle called the proteasome. Under inflammatory conditions the proteasome's structure is altered so that it more efficiently produces peptides that associate with the MHC-Class-I peptide carriers and the resulting complexes are displayed on the cell surface

[23]. This spectrum of peptide-MHC-Class-I complexes on the cell surface thus provides a snapshot of what is currently going on inside that cell. Remember, this is happening in every nucleated cell in the body, so if you think that this sounds like a very large investment of energy and metabolites, then you are right. It is.

Peptide-MHC-Class-II complexes, in contrast, are not expressed on all cells, but rather are restricted to the so-called Antigen Presenting Cells (APCs) such as the phagocytic dendritic cells or macrophages, or the B-cells that internalise antigens via their BCR. These APCs take up material from their surroundings into endosomes where the ingested proteins are digested to peptides. These peptides may then associate in the endosome with MHC-Class-II molecules, and the resulting complexes are displayed on the APC's surface. These peptide-MHC-Class-II complexes thus provide a snapshot of what an APC has recently eaten.

Each individual human can express 3 to 6 different MHC-Class-I and 4 to 8 different MHC-Class-II molecules. Unlike the receptors of innate immunity or the adaptive immune receptors of B-cells and T-cells, all of which are highly specific for particular ligands, the MHC molecules have a rather relaxed binding specificity. Thus, though each MHC-Class-I molecule has its particular peptide-binding preferences, nevertheless each of them is thought to be able to bind around a million different peptides with useful affinity [24]. However, given that there are a strictly limited number of peptide-binding MHC molecules per haploid human genome, a million is not a large number, for it implies that each of us can, at best, bind roughly 3 to 6×10^6 different peptides on MHC-Class-I molecules. That is not a lot, given that the total possible array of nonamer peptide sequences is around 5×10^{11}. For the many species that express only one or two MHC-Class-1 molecules, the situation is even worse. So why don't we all have many more peptide-binding MHC molecules? Why not scores, hundreds or even thousands of them? The answer is that, in a sense, we do, for the MHC locus is highly polymorphic and thousands of different alleles are spread through the population. That tells us that lots of different MHC-Class-I and Class-II alleles are a good idea, and that there is, in fact, no problem in making lots of different MHC structures. So why the restriction to just a handful in each individual? Perhaps the answer is simply that MHC-Class-I molecules are expressed in every nucleated cell and MHC-Class-II molecules in every APC as well as in certain other cells such as, for example, thymic epithelial cells (see Sect. 4.8). This represents a massive investment of energy and metabolites. Because of this, the number of MHC pseudo-alleles present in an individual will be a trade-off between the investment of metabolic resources needed to express them and the fitness advantage gained from having a peptide-recognising T-cell system.

So why two types of peptide carrier molecules? The simple answer is that peptide-MHC-Class-I complexes on the surface of a cell provide information about "non-self", in the form of pathogens that may be lurking inside a cell. The peptide-MHC-Class-II complexes, in contrast, provide information about "non-self" taken up from the immediate environment of the cell.

4.6.3   TCRs Recognise Complexes of Peptides and MHC Molecules

The two quite different types of information, provided by the Class-I and Class-II peptide complexes, are processed by two different types of $\alpha\beta$ T-cells. In some $\alpha\beta$ T-cells the TCR on the surface works hand in hand with a co-receptor molecule known as CD4. These CD4$^+$ $\alpha\beta$ T-cells recognise peptide-MHC-Class-II complexes present on APCs, and hence they are in the business of recognising peptides derived from proteins that were digested in the endosome. A T-cell that instead expresses the co-receptor CD8 interacts instead with peptide-MHC-Class-I complexes, and thus is in the business of detecting peptides that were generated by digestion of intracellular proteins in the proteasome. Proteins in the endosome do not, in general, have ready access to the cytosol and hence will not be efficiently presented on MHC-Class-I molecules. On the other hand, proteins in the cytosol do not, in general, have ready access to the endosome, and hence will not be efficiently presented on MHC-Class-II molecules.

The fact that the TCR recognises a complex of peptide with an MHC-molecule rather than just a naked peptide has enormous consequences for the way adaptive immunity functions. In the B-cell system a huge—for all practical purposes—infinite array of possible ligands is met, and countered, with an enormous ($>10^{14}$) potential repertoire of different BCRs. In the T-cell system things are very different. The RAG-dependent rearrangement of the TCR genes is thought to be able to produce a repertoire of over 10^{14} different receptors. An array of something like 5×10^{11} different nonamer peptide ligands on the one side, being faced by 10^{14} receptors on the other, would sound not unreasonable. That, however, is not the way it really is, for the ligands that the TCRs recognise are not naked peptides but rather peptide-MHC complexes. Thus, when a TCR recognises a peptide-MHC complex, part of the binding energy comes from interactions between the TCR and the MHC molecule, and part from interactions of the TCR with the peptide. Since the number of MHC molecules available to an individual is strictly limited (Sect. 4.6.2), and each of these binds only a limited range of different peptides, the number of different peptide-MHC complexes that an individual can present to T-cells is also strictly limited.

4.7   Which TCRs Are Potentially Useful: "Positive Selection"

A TCR that cannot productively interact with any of the MHC-Class-I or Class-II molecules available in an individual is a TCR, which is of no value to that person's immune system. Flooding the immune compartment with huge numbers of T-cells that are incapable of recognising any of the available peptide-MHC-Class-I complexes would clog up the system, and so all the "useless" cells must be got rid of

[25]. This process, which only retains the potentially useful receptors, takes place in the thymic cortex and is known as "positive selection". It is believed to involve some means of measuring the strength of the association of the T-cell's TCR with MHC-peptide complexes expressed on the surface of the cortical epithelial cells. Those T-cells that fail to interact productively with any of these complexes are directed into apoptosis and die. The question of the number of different MHC genes per genome is thus crucially important here. Natural selection has had to find the best possible compromise between the energy drain involved in expressing MHC molecules in every cell in the body, the waste involved in killing large numbers of T-cells whose receptors do not "fit" to any of the individual's MHC molecules and the fitness benefit derived from having T-cells able to look into what is going on inside other cells.

4.8   "Negative Selection" of T-Cells in the Thymus

Perhaps the greatest challenge in considering the evolutionary origin of adaptive immunity in vertebrates revolves around the simple chicken-and-egg problem that somatic recombination of receptors cannot arise in the absence of appropriate tolerance systems, and tolerance cannot evolve in the absence of the repertoire. It is hard to envisage two such complex systems as repertoire generation and tolerance induction suddenly evolving in tandem. However, the absolute necessity for tolerance systems is underscored by the investment that is made to purge the repertoire of anti-"self" specificities.

Positive selection in the thymus cortex of cells expressing TCRs that will interact with one of the individual's MHC molecules involves a massacre of the newly formed T-cells—but this is just the start. What follows is a second massacre—one that takes place in the thymic medulla and is known as negative selection. What is the purpose of negative selection? Why is it needed? How does it work? The function of the repertoire of TCRs on mature CD8$^+$ T-cells is to continually scan the peptide-MHC-Class-I complexes exposed on the surface of normal cells. Those cells, which display only peptides derived from normal "self" proteins, must be left in peace. Those, which display peptides derived from the proteins of an intracellular pathogen, must be destroyed. How are the CD8$^+$ T-cells to distinguish "self" peptides from "pathogen" derived ones? The answer is that in reality they can't— and they don't have to, because those CD8$^+$ T-cells that can recognise "self" peptide-MHC-Class-I complexes are destroyed in the thymus by the process of negative selection [25].

In a similar way the CD4$^+$ T-cells are there to recognise peptide-MHC complexes displayed on the surface of APCs. These complexes are derived from material that the APC has taken up from its surroundings, and processed to peptides in the endosomes. The resulting peptides may be derived from pathogens, like bacteria or viruses, or they may be derived from "self" proteins present in apoptotic cells or in the debris from virus-infected cells that the APC has scavenged. All of this material will be processed in the endosomes, and thus not only pathogen derived but also "self" peptides will be presented by the APC on MHC-Class-II molecules to the CD4$^+$ T-cells. How are the CD4$^+$ T-cells to recognise the "non-self" components, and yet ignore the "self" components? As with the CD8$^+$ T-cells, they can't, and as with the CD8$^+$ T-cells they don't have to, because those CD4$^+$ T-cells that recognise "self" peptide-MHC-Class-II complexes are destroyed in the thymus medulla by negative selection [25].

Negative selection works by having the thymic medullary epithelial cells present "self" peptide-MHC-Class-I complexes and "self" peptide-MHC-Class-II complexes to those T-cells that emerge from positive selection in the cortex. T-cells, which interact with high affinity with any of the "self" peptide-MHC complexes, are driven into apoptosis and die. For this selection to function properly, the medullary thymic epithelial cells face two major problems. The first is that if negative selection is to be effective, then these epithelial cells must be able to present all "self" peptides to the T-cells. The second is that they must be able to present "self" peptides not only on MHC-Class-I to CD8$^+$ T-cells but also on MHC-Class-II molecules to CD4$^+$ T-cells.

4.8.1 Presenting All "Self" Peptides

Thymic epithelial cells are end-differentiated cells, and end-differentiated cells express only a specific subset of all the genes in the genome. Because of this, the expression of many genes is restricted to certain tissues, and so one would not expect that epithelial cells in the thymus to be able to express the tissue restricted genes typical of other tissues. In this respect, however, the medullary thymic epithelial cells are different for, they express a gene called AIRE. This gene codes for a protein, which, in ways that are not yet fully understood, licences these cells to express essentially every gene in the genome. Not all genes are expressed in every medullary thymus epithelial cell all of the time, but at any given moment a tissue restricted protein is expressed by roughly 1–3% of them. These proteins, synthesised in the medullary epithelial cells, are then processed through the proteasome and presented as peptide-MHC-Class-I complexes to the CD8$^+$ T-cells. This allows for the detection and destruction of those newly made CD8$^+$ T-cells that express autoimmune receptors [25].

4.8.2 Expressing "Self" Peptides on MHC Class-II

The peptides detected by CD4$^+$ T-cells are different from those detected by CD8$^+$ T-cells. This is because peptides formed in the endosome are generated by a different set of proteases from those that are present in the proteasome. In presenting these "self" peptides to CD4$^+$ cells the medullary epithelial cells must overcome a peculiar problem. Normal cellular proteins can be readily routed to the proteasome, and the resulting peptides can be loaded onto MHC-Class-I molecules. However, these normal cellular proteins do not usually have access to the endosomes. Only proteins from phagocytosed material are directed to the endosome. How then can the medullary epithelial cells make "endosome type" peptides from these normal proteins, and present them as MHC-Class-II complexes to $CD4^+$ $\alpha\beta$ T-cells? The answer seems to be that these medullary epithelial cells, as well as the medullary dendritic cells, are peculiarly efficient at autophagy—the mechanism of "internal phagocytosis" by which normal cellular proteins are transferred to the endosomes (see Sect. 2.2.5). In this way samples of the cell's own proteins reach the endosomes where they are digested to peptides that associate with MHC-Class-II molecules.

4.9   The Price of T-Cell Selection in the Thymus

The result of positive and negative selection in the thymus is that the $CD4^+$ and the $CD8^+$ T-cells, which emerge into the periphery, carry receptors that ought to be able to make use of an MHC molecule present in that individual, and they ought to be incapable of interacting with any "self"-peptide-MHC molecules. The repertoire of receptors on these T-cell populations will, nevertheless, almost certainly contain specificities that will detect peptide-MHC complexes containing peptides derived from pathogens. This is what makes the $CD8^+$ and the $CD4^+$ repertoires selectively advantageous. The price, however, is enormous, for over 95% of the T-cells die during positive and negative selection in the thymus, and only the few survivors enter the circulation. Nevertheless, in a young human being, around 10^9 survivors per day do exit the thymus [26]. The daily destruction of billions of cells in the thymus is an extraordinary waste of metabolites and energy. Comparable size-adjusted numbers of destroyed thymocytes have been reported for other vertebrates such as mice and chickens.

4.10   Co-evolution of TCRs and MHC-Molecules

MHC molecules, by displaying peptides on cell surfaces, provide the T-cells with information about what is going on inside other cells—this is the sole established function of MHC molecules. Attempts to find some alternative function of these MHC molecules, which would provide them with a selective advantage in the absence of an adaptive immune repertoire, have not so far been successful. T-cells interpret the information provided by the peptide-MHC-complexes—this is the sole function of the TCR. MHC molecules and TCRs thus represent an extraordinary case of co-evolution, and it is unclear how either of them could have arisen on its own, for $\alpha\beta$ T-cells, as we know them today, make no sense in the absence of MHC molecules, and MHC molecules make no sense in the absence of $\alpha\beta$ T-cells. How these two molecular species, with their very different, but complementary, functions, could have co-evolved is a mystery that has puzzled biologists for decades. One clue may be that MHC molecules share with the TCR the use of the C1-type IgSF domain, which is restricted to gnathostomes, and this suggests that the genesis of the TCR and MHC families was not totally independent. What is clear is that the peptide-binding MHC molecules emerged in parallel with RAG-mediated recombination, as the jawed vertebrates split from the earlier non-jawed vertebrate line [27]. However, it is certainly possible that the "original T-cells" were more like B-cells in that they did not recognise peptides, and so did not require the MHC peptide-binding molecules. The reason for thinking this is that in addition to the αβ T-cells, gnathostomes also possess a second population of T-cells which express a TCR composed of one γ and one δ chain. These γ and δ chains are homologous to the α and β TCR chains of "conventional" T-cells. The "gamma-delta" T-cells are thought to play roles in mucosal immunity and they seem to bind intact proteins rather than peptides, and hence do not require MHC for antigen recognition. The gamma-delta T-cells may represent the vestiges of a T-cell population that preceded the evolution of the αβ T-cells. Were this to be the case, then the evolution of the TCR may well have preceded that of the MHC-Class-I and Class-II molecules.

4.11   Repertoire Change After Central Tolerance: Somatic Hypermutation of BCRs

One might expect that after central tolerance has been established, there would be an absolute prohibition on any subsequent widespread mutation of the sequences coding for the antigen-binding site, for this might well give rise to new autoimmune specificities. This expectation is fulfilled for mammalian T-cells, but for the B-cells things are different.

The binding sites of adaptive immune receptors are formed from the three "Complementarity Determining Regions" which contact the ligand (Fig. 4.7). Since RAG recombination only mutates the sequence of CDR3, but leaves the sequences of CDR1 and CDR2 untouched, the repertoire that is produced is far from being as complete as it might be. Because only a small fraction of an incomplete B-cell repertoire can be expressed at any one time, it is unlikely that a pathogen entering the body will be met by B-cells expressing the "perfect" receptor against it. Nevertheless, the available repertoire is large enough that, in general, some B-cells with less than perfect—and hence low affinity—receptors will be available. This handful of cells would never be able to effectively counter a rapidly growing bacterial or viral pathogen, and so after contact with the pathogen, they are directed to specialised niches in the secondary lymphoid tissues where germinal centres develop, a micro-environment required for the "maturation" of the immune response. In the germinal centres the activated B-cells undergo rapidly accelerated cell divisions and, multiple rounds of somatic mutation, followed by a somatic selection process to filter out the cells with the highest affinity receptors. The

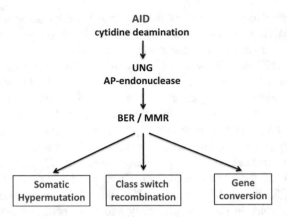

Fig. 4.10 Induction of mutations by Activation-Induced Deaminase (AID). Deamination of dC by AID to form dU is followed by the removal of the uracil moiety by Uracil-DNA-glycosylase (UNG) and the introduction of a nick in the DNA by the AP-endonuclease. Different elements of the Base Excision Repair (BER) or of the Mismatch Repair (MMR) systems are recruited to repair the damaged DNA strand and may result in somatic hypermutation, class switch recombination and/or gene conversion [29]

value of this can be seen by the fact that cold-blooded vertebrates, which lack germinal centres, have problems "maturing" their immune responses [28].

For this process of "somatic hypermutation" in the germinal centres, RAG recombinase is of no value for its mutagenic action is part and parcel of the rearrangement of the V, (D) and J sequence modules, which was completed before the B-cell emerged into the periphery. Instead a different mutagenic mechanism—one centred on the cytidine deaminase AID—comes into play [29]. AID is the gnathostome homolog of the cytidine deaminases used by agnathans to initiate the gene conversion process by which their adaptive immune receptor repertoires are generated (Sect. 4.3.1). The formation of dU by deamination of dC is a mutational process that is repaired by the cell's DNA repair systems. These repair systems are very ancient and have been naturally selected for their ability to restore the original sequence after a mutational event. It is at the moment unclear what induces these repair systems to instead generate mutations in the context of B-cell somatic hypermutation, class switch recombination and in gene conversion of V-genes in chickens, rabbits and sheep (Fig. 4.10). It seems, however, that the type of mutation that is induced depends on the nature of the cofactors in the cell that work together with AID.

During the AID-initiated somatic hypermutation process, mutations are introduced into the rearranged V-gene sequence and thus can alter not just CDR3, but CDR1 and CDR2 as well [30]. This somatic hypermutation process is fraught with problems. Not only will the mutation process result in B-cells with structurally defective receptors but, in addition, the tiny fraction of mutant cells with increased affinity for the antigen must be reliably identified and their numbers quickly expanded in the germinal centres. Selection and expansion are achieved by making the cells compete with each other for access to a limited amount of antigen, which is presented in intact form on the surface of the germinal centre's "follicular dendritic cells". Mutant B-cells with low affinity receptors lose out in this competition for antigen, and, having been left empty handed, they are condemned to death by apoptosis. However, those cells with a higher affinity receptor grab the antigen, internalise it, process it to peptides in their endosomes and express the resulting peptides as MHC-Class-II complexes on the cell surface. CD4$^+$ "follicular T-helper cells" that recognise these peptides will then provide the B-cell with essential survival factors. These high affinity winners may differentiate into effector cells or go through new rounds of division, mutation and selection, so that in the space of a few days clones of cells expressing high affinity receptors have been selected and expanded [31]. It is, however, clear that changing the structure of the CDRs can dramatically alter not only the affinity of the BCR for the cognate ligand, it may also radically change its ligand specificity, and thus may generate an autoimmune receptor. The central tolerance mechanisms operating in the bone marrow have been left far behind, so other means of dealing with B-lineage cells that have acquired potentially autoimmune receptors must be available.

4.12   Life After Central Tolerance: Peripheral Tolerance and Lymphocyte Activation

Experience teaches us that nothing in life is perfect, and this applies also to central tolerance. Though the T-cell compartment does not do anything as bizarre as mutating its receptors after central tolerance is established, that does not mean that the repertoire is completely safe. All TCRs must be at least a little bit autoimmune, because they must interact with "self" MHC molecules and so "self" and "non-self", "safe" and "autoimmune" are neither well defined nor clearly distinguished categories. The highest affinity receptors are potentially "autoimmune", and the lowest are "safe", but there is a considerable grey area between "low" and "high", and no simple affinity cut-off above which cells must be eliminated and below which they are perfectly safe. In consequence, many of the B- and T-cells entering the periphery express receptors that are somewhere between "borderline autoimmune" and "frankly autoimmune". This should not just be viewed merely as a failure of central tolerance, for it is true that "*Immunity operates on the edge of autoimmunity. The more potent an immune response is, the greater the risk of auto-reactivity and self-harm*" [32]. In addition to the problems posed by the question of an appropriate affinity cut-off for B-cells and T-cells, there is in the T-cell compartment, a further problem. This is that the astonishing system of AIRE-mediated, global gene expression in thymic medullary epithelial cells, is not as all-encompassing as one might think. Tissue-specific splice variants or tissue-specific post-translational modification of protein structure will not be covered and thus T-cells cannot be centrally tolerised to these "self" structures. Simply allowing centrally tolerised lymphocytes to do their thing unchecked in the periphery would result in a catastrophe [25]. Somehow such potentially autoimmune cells must be controlled. An intricate set of overlapping controls has evolved to do this, and their importance is seen in the devastating autoimmune diseases that appear should these systems fail. It is also seen when pathogens—such as tumours—manipulate these regulatory systems to their own ends. These controls fall roughly into two groups—those based on the "fail-safe two key" strategy, and those due to the presence of regulator cells.

4.12.1   The "Two Key" Fail-Safe Strategy

Innate immune cells, such as macrophages, express many different receptors, and these together provide a "picture" of a potentially dangerous situation. Each of these cells is empowered to analyse the input signals flowing in from its receptor repertoire, and then autonomously decide what should be done. These cell autonomous decisions work, because the receptors, which are providing the input data, have been honed over millions of years of selection to a fine state of tolerance. These receptors can therefore, in general, be trusted. The situation is different with lymphocytes expressing somatically evolved immune receptors. Here the receptors have been screened by the rough and ready business of central tolerance and they cannot be relied on to be truly tolerant. Because of this, in adaptive immunity cell autonomous decisions are avoided like the plague. Instead the "two key" principle guiding the launching of nuclear weapons applies, so that at least two different cell types must agree that there is a problem that requires firm action before an adaptive immune response can be initiated.

One can see this schematically in Fig. 4.11, which shows the activation of a CD8$^+$ killer cell. A brief description of this complex process is given in the figure legend, but the important take-home lessons are actually very simple. First, the activation of a CD8$^+$ T-cell requires the collaboration of three different cell types and involves five different ligand-receptor repertoires. Furthermore, both the CD4$^+$ and CD8$^+$ T-cells require additional information from the dendritic cell in the form of "co-stimulation", and the CD8$^+$ T-cell also requires "help" from the CD4$^+$ T-cell [33].

Why does it have to be so complicated? The answer is that neither the CD4$^+$ nor the CD8$^+$ TCR repertoires are entirely trustworthy. Only if both of them, together with the dendritic cell, agree that there is a problem that needs to be addressed, will a clone of activated CD8$^+$ killer cells be formed. This caution is well justified because once the activated CD8$^+$ T-cells have been formed they need no help or permission from any other cell to go about doing their business. Their business is killing. A CD8$^+$ T-cell accidentally activated against a "self" peptide will kill any cell that expresses the appropriate peptide-MHC-Class-I complex on its surface—and that is the worst possible news, because CD8$^+$ T-cells are serial killers. The activation scheme shown in Fig. 4.11 is set up so that if any of the signals are missing, then the activation is aborted. When the system fails, it ought to fail "safe".

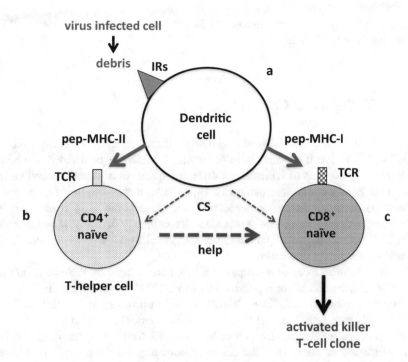

Fig. 4.11 Activation of a naïve CD8$^+$ T-cell. This requires three different cell types—a dendritic cell, a CD4$^+$ "helper" T-cell and the CD8$^+$ T-cell. It involves a total of five different immune receptor repertoires: the innate receptors on the dendritic cell; the peptide-MHC-Class-I complexes (pep-MHC-I) displayed on the dendritic cell, the peptide-MHC-Class-II complexes displayed on the dendritic cell, the CD4$^+$ T-cell receptor repertoire and finally the CD8$^+$ T-cell receptor repertoire. A dendritic cell (a) detects the danger signals associated with the debris of a lysed virus-infected cell by way of its innate immune receptors (IRs) The debris is phagocytosed and peptides generated from it by digestion in the endosome are displayed on the cell surface as peptide-MHC-Class-II complexes (p-MHC-II) to the T-cell receptor (TCR) of a naïve CD4$^+$ T-cell (b). The dendritic cells also carry on their surface the so-called co-stimulator (CS) molecules (narrow stippled arrow), with which they can identify themselves to the CD4$^+$ T-lymphocyte as officially approved "Antigen Presenting Cells". These dendritic cells are also adept at "cross presentation" by which proteins in the ingested material are transferred into the cytosol, digested in the proteasome and displayed on the cell surface as peptide-MHC-Class-I (pep-MHC-I) complexes to CD8$^+$ T-cells (c). If a naïve CD8$^+$ T-cell interacts with the peptide-MHC-Class-I complex, and if it is also assured by the co-stimulatory signals on the surface of the DC that this is indeed a bona fide "Antigen Presenting Cell", and if it also receives appropriate signals as from the CD4$^+$ helper cell (thick stippled arrow), then the naïve CD8$^+$ T-cell will start to divide and produce a large clone of activated "killer cells"

It is not only CD8$^+$ T-killer cells whose activation is regulated in this way, for a similar sort of control regulates the activation of B-cells. Not all autoimmune B-cells are removed by central tolerance in the bone marrow, and some do make it out into the periphery. There more may be generated during the process of somatic hypermutation in the germinal centres. B-lymphocytes, however, are normally not licenced to autonomously mount a response to their cognate ligand. They require help from T-cells to initiate an immune response. If a B-cell meets an antigen in the periphery, and fails to get the appropriate T-cell help, then it will be converted into an inactive "anergic" state.

4.12.2   T-Regulator Cells

Within the CD4$^+$ and CD8$^+$ T-cell populations there are many functional subdivisions. The CD4$^+$ T-cell population, in particular, is known to be divided into a large number of subsets each of which has a different spectrum of functions and each of which is defined by a typical pattern of expression of transcription factors within it. These transcription factors do not alter the structure of the TCR in any way, but, by deciding which genes will be expressed in the cell, they define how an interaction of the TCR with its cognate ligand will be interpreted, and hence what functions the cell will subsequently undertake.

Perhaps the most extreme example of a function dictated by the expression of a particular transcription factor is provided by those CD4$^+$ T-cells whose TCR fell into the grey area between "safe" and "autoimmune" during negative selection. These CD4$^+$ T-cells may switch on the transcription factor FoxP3, which turns them into "T-regulator" cells (T-regs). In such cells the signal from engagement of the TCR with its ligand does not instruct the cell to provide help for the activation of B-cells or of CD8$^+$ killer cells, but instead T-regs do exactly the opposite—they switch the activation of other cells off.

These different T-cell lineages are not cast in concrete, and they may, under certain circumstances, demonstrate a degree of plasticity. This is exploited by many cancers, which are able to subvert the CD4$^+$ helper cells by turning them into "induced T-regs". The adaptive immune response to the tumour then grinds to a halt. Learning how to manipulate the plasticity of CD4$^+$ T-cell populations will have obvious clinical relevance.

4.13   Beyond the Receptor Repertoire: Lymphocyte Effector Functions

So far we have concentrated on the mechanisms that generate an adaptive immune receptor repertoire, and on the means, which have evolved to purge the repertoire of autoimmune specificities. Only once these processes have been completed can the repertoire be employed to counter pathogen infections. Immune defence requires not just sensors to detect a problem; it also needs effector systems to remedy it.

The MHC-Class-I-dependent CD8$^+$ killer T-cells provide a means to destroy virus-infected cells. These killer cells, once activated, destroy their targets by directly transferring the contents of cytotoxic granules into the target, which then dies by apoptosis. Killer cells appear to have arisen early in vertebrate evolution, for there is genomic sequence evidence indicating that they are already present in the surviving basal cartilaginous fish [27].

The effector functions of gnathostome B-cells are somewhat more complex. During the initial interactions with a pathogen, the B-cell collects information on the cytokines and other factors in its immediate environment and carries out a computation that provides a decision as to what effector system should be linked to the antigen-binding part of a BCR. In some cases it may be best to ensure that the receptor will activate the complement system, in others that the pathogen be phago-cytosed by macrophages, in other cases it may be better to have the receptor secreted across a mucosal surface, while in yet others it may be important that the receptor can cross the placenta. Each of these responses is best achieved by treating the entire antigen-binding part of the BCR as a modular unit that can be plugged onto one of a set of different "constant" regions that define the effector function of the molecule. The process that makes this possible is known as "class switch recombination". It involves yet another recombinational process at the DNA level. In this case the entire V-domain, with its three CDRs, is physically transferred to one of a number of different C1-type domains that enable different functions of the antibody molecule [29]. This recombination process is initiated by the enzyme AID, which is also central to somatic hypermutation, and whose homolog in agnathans is required for initiating the gene conversion process involved in repertoire generation (see Figs. 4.2 and 4.10). A sophisticated form of class switch recombination is already present in amphibians and all later vertebrates, but a simplified version of this process is evident in the earliest jawed vertebrates currently available—the cartilaginous fish [34].

Activation of B-cells via the BCR may lead to differentiation of the B-cell into a plasma cell that synthesises large amounts of the receptor and secretes it as soluble antibody at a rate of around 2000 molecules per second. To be secreted the receptor has to be modified so that it is no longer bound to the cell surface. In principle this change, like V, (D), J rearrangement or like class switch recombination, could be effected by recombination at the DNA level, but there is no law which says that evolution has to be consistent, and the solution that emerged involves removing the exon containing the transmembrane domain by directed alternative splicing at the RNA level.

4.14 Adaptive Memory in Gnathostomes

If you have just been infected with a novel pathogen, then clonally selected B and T-cells are not going to provide significant defence for the first week to 10 days. Over this crucial initial period those who have not been vaccinated against the pathogen will have to rely on innate immunity to survive. This underscores that the selective advantage provided by adaptive immunity is probably not its slow initial response, but rather the fact that it remembers the pathogens that it has come in contact with. This is because activated T and B-cells are set aside during an initial response and retained as the so-called memory cells.

In the B-cell compartment this memory response is made up of two arms. The first of these consists of antibody secreting plasma cells that are produced after activated B-cells have been selected in germinal centres, and matured by somatic hypermutation and class switching. Once an infection has been overcome these plasma cells may migrate to the bone marrow where a small fraction of them settle down and continue to produce antibody for long periods of time—sometimes for decades [35]. The second arm of memory is provided by antigen stimulated B and T lymphocytes that differentiate into circulating memory cells. Should the pathogen reappear in the future, then circulating antibody provided by the long lived plasma cells will make life hard for it, and at the same time the memory B- and T-cells will be driven into cell division to produce a whole new generation of plasma and memory cells.

Thus, as one grows up in a particular environment, one gradually acquires immunological memory directed against all of the endemic pathogens. Over this initial period the young animal is at risk, but after a while immunological memory provides protection that will effectively neutralise an infection at the earliest stages—often even before symptoms develop. B-cell adaptive immunity is thus a large-scale exercise in vaccination. These vaccinations provide an enormous selective advantage, and they are what justify the enormous costs involved in developing specific adaptive immune responses.

4.15   Diversity of Adaptive Immune Repertoire Formation in Gnathostomes

Across the phylogenetic span from nematode worms to mammals, immune defence involves the use of a mix of "innate", i.e. germ line encoded, and of "adaptive", i.e. somatically encoded, receptors. Since different species inhabit different environments, it is clear that neither innate nor adaptive immunity can be viewed as an "off the peg" defence system. Though natural selection has ensured that certain features of innate immunity are common between the fruit fly and mammals, there are also very considerable differences. In the same way, and for the same reasons, the details of the workings of adaptive immune defence varies considerably between different gnathostome species.

The brief outline given here of the gnathostome adaptive immune receptors is heavily skewed to the situation in mice and humans, for these are by far the best-studied vertebrates. If one broadens the perspective, to cover other species, then it turns out that there is little in adaptive immunity that is precisely conserved across all jawed vertebrates. This is perhaps not surprising for no adaptive immune system is ever truly "complete". New pathogens with new virulence strategies constantly require adjustments to the current immune system, and since different vertebrate species live in very different environments, and face different spectra of pathogens, it would be surprising indeed if there were no differences in the ways their immune systems work. Indeed when the adaptive immune response in any species is examined in detail a host of idiosyncratic features emerge, and these make the application of the results of animal experiments to human clinical medicine an enterprise that is often fraught with uncertainties. True, the division of the lymphocyte universe into T- and B-cells whose receptors are generated by a process involving RAG recombination are general features in gnathostomes. Beyond that, however, divergence is the name of the game. A few examples will demonstrate this.

The first example we will take is that the RAG catalysed VDJ recombination system is not entirely sacrosanct. RAG requires multiple alternative V, (D) and J modules with which to generate the repertoire but, on the other hand, there is a general problem with gene families in which the various members all carry out very similar functions. As such a family increases in size, selection pressure on each member drops and their maintenance by purifying selection becomes increasingly difficult. The genes start to accumulate random mutations and degrade to pseudo-genes. For example, in the first human heavy chain locus that was completely sequenced there are 123 V-gene segments, but 79 of these are pseudo-genes, so that there are only 44 intact V-gene segments left that can be used [36]. In a number of vertebrates including the chicken and rabbit this process has gone one step further, for there is just one single V-gene segment left functionally intact. In these species the problem of having too few V_H-gene segments is solved by taking a leaf out of the lamprey's book: the last remaining intact V-gene segment is rearranged to D and J using RAG recombinase, after which AID-mediated gene conversion is used to copy information from the pseudo-genes into the rearranged VDJ-gene segment, and so produce a large and diverse repertoire.

A second example of variation of the adaptive system in gnathostomes is that the antigen-binding site of a BCR does not always consist of two polypeptide chains, as described in Sect. 4.4.3 (Fig. 4.7). There are bizarre forms, like the IgNAR of sharks, in which the antigen-binding site consists of a single V-type domain. The group of camels, dromedaries and llamas separately evolved a rather similar sort of molecule, in which the binding site is formed by a single V-domain, and hence like the shark IgNAR, contains only 3 CDRs. This form of camel immunoglobulin has excited interest in recent years because the isolated variable domain binds antigen, and it is smaller, more soluble, more stable and much easier to produce than similar structures from conventional antibodies. These camel V-domains are of potential pharmacological value and may also be of use for research purposes to provide small, but specific, probes that can be expressed inside eukaryotic cells.

A third example of variation concerns the $CD4^+$ T-helper and the T-regulator cells, which in humans and mice are the central controllers of immune responses, and yet in certain bony fish, including the Atlantic cod, the CD4 co-receptor and the MHC-Class-II molecules have been lost [37]. Since other bony fish do have an intact CD4 system, it seems that the cod's loss of CD4 is a derived characteristic. In this particular case an expansion of MHC-Class-1 pseudo-alleles suggests that the loss of the CD4 compartment may be compensated to some extent by an expansion of the role of $CD8^+$ T-cells. In a similar vein, zebra fish that have been genetically altered so that they are unable to form B- and T-cells are not obviously disadvantaged, at least under laboratory conditions [38], while the same mutation in mice or humans results in a frequently lethal "Severe Combined Immune Deficiency". Perhaps the aqueous environment is less packed with pathogens than is ours, so that fish can afford to lose bits of their immune system in ways that we cannot.

4.16   The Evolutionary Relationship of the Agnathan and Gnathostome Adaptive Immune Systems

When one looks at the range of solutions to life's problems that living systems have come up with, then it is fair to conclude that there is no such thing as a problem to which there is only one possible solution. The politicians' favoured gambit, "there is no alternative", does not apply in biology. However, as François Jacob pointed out, the range of alternatives that are available are indeed constrained by one important factor—history. A simple system that has not yet developed in any very sophisticated fashion has the freedom to evolve in many different ways, so that when it seeks a solution to some problem it may have access to a large battery of options. However, the number of evolutionary paths open to a complex system is reduced. As a result, two unrelated complex systems may reach similar analogous solutions to a problem, simply because their levels of complexity have narrowed the range of evolutionary options available to them in a remarkably similar way. The problem therefore is that seemingly similar solutions can be reached by very different means, and, to make matters worse, sometimes seemingly different solutions can in reality be closely related at the genetic level.

"Homology" is a technical term that implies inheritance by descent. A character in two animals is homologous if it is uniquely dependent on shared inherited genes or genetic circuits that were present in the most recent common ancestor. The idea of shared genetic information is important to keep in mind, for it is the key to distinguishing homology from analogy. Analogy describes a situation in which two different species have reached similar solutions to a problem by convergent evolution. Analogy implies no relatedness in terms of descent. However, analogy is much more than just "not homology", for it tells something about the selective forces which were at work.

When one compares the adaptive immune systems of agnathans and gnathostomes, then it is clear that there are fundamental differences between them. It is equally clear that there are a number of astonishing similarities. What are these similar features that must have been inherited from their most recent common ancestor? What are the divergent features that each of them added on since they diverged from that most recent common ancestor? These questions are not as easy to answer as one might think, for while some homologies are informative, others are—in this context—entirely trivial. Even worse, as we will see in the next section, some analogies may be based on "deep" homology (see also Appendix E).

4.16.1 Homology and Analogy

What we would like to know is how adaptive immunity came to be, and why it developed along two rather different lines, one in agnathans and the other in gnathostomes. The only tools we have available to tease apart what happened is to ask if homologies exist between the agnathan and gnathostome systems, and, if such homologies can be identified, then see what they can tell us about the last common ancestor of all vertebrates, and the evolution of adaptive immunity in agnathans and gnathostomes. When one looks at the antigen-binding receptors—VLRs in the case of agnathans, IgSF domain-based receptors in the case of gnathostomes—then it seems clear that they are very different. The same is true if one considers the genetic rearrangement processes employed to generate the receptor repertoires—gene conversion in agnathans, and RAG driven recombination in gnathostomes. However, on the other hand, there are many features, which make the two systems seem very similar indeed.

Which, if any, of the similarities are real evidence of homology, i.e. of shared inheritance by descent? For example, both agnathan and gnathostome systems use antigen-specific receptors—VLRB in agnathans, BCR in gnathostomes—first as cell surface bound molecules, and then as released effectors. Is this striking shared character evidence of homology? In the gnathostome case, the receptor is held on the surface by virtue of a membrane-spanning domain. Releasing the receptor as a soluble molecule is achieved by alternative splicing of the messenger RNA to remove the membrane spanning exon. In the agnathans, on the other hand, the VLRB receptor molecule is bound to the surface of the cell by means of a post-translational modification that adds a so-called GPI-anchor onto the protein chain. This anchor links the receptor to the cell membrane. Omitting this post-translational modification permits secretion of the product. The two systems thus use different molecular means of linking the receptor to the membrane, and different ways of releasing the cell-bound receptor in soluble form. The result may look broadly similar, but this is a case of convergent evolution—of analogy rather than of homology.

Even in cases where a character in agnathans is clearly homologous to one present in gnathostomes, because it results from inheritance by descent, this may not necessarily be terribly informative. To take a simple example, agnathans and gnathostomes—like all other life forms—use a triplet code to convert the information encoded in their genome into the amino acid sequences of proteins. This is certainly a shared, inherited, homologous feature—but since there are no exceptions, it is, for our purposes, trivial. By the same token, any phenotypic characteristic, which is a direct consequence of the use of such a triplet code, is also, for our purposes, trivial and uninformative. A triplet code has the consequence that random mutation of the coding region of a gene will frequently disrespect the "rule of three" and hence will generate large numbers of frame shift mutations (Sect. 4.5.1). Since agnathans and gnathostomes both use random mutation mechanisms to alter the coding sequences of their adaptive immune receptors, it is inevitable that in both cases a great deal of junk will be formed. In both cases some means must be found to select and preserve the well-formed receptors [39]. The mere fact that selection is necessary, and takes place, in both systems is thus an inevitable consequence of the use of a three letter code. The question of homology then rests on whether some unusual process of selection of the products is shared by the two systems. Though a great deal is known about selection in gnathostomes, at the moment almost nothing is known about how this is achieved in the agnathan system. Until we have that knowledge, the question of whether selection of the products in agnathans and gnathostomes is an example of homology or merely a consequence of the universal use of a triplet code, remains open.

Other features of agnathan and gnathostome adaptive immunity that appear at first sight to be evidence of striking homologies turn out on closer examination to be less convincing. For example, both systems make use of a large collection of lymphocytes each of which expresses just one single receptor specificity. Is this remarkable similarity a case of homology? The idea that these two independent systems would reach the same solution to the problem of expressing a very large repertoire by using a "one lymphocyte, one specificity" solution might seem unlikely, but the degree of unlikeliness has to be measured in terms of what other solutions are available. That is, of course, *a priori* impossible to judge, but what one can say is that any solution would have to be compatible both with the way that fundamental molecular processes of cell biology have evolved, and with the number of receptors that have to be accommodated. Alternative explanations, from Ehrlich's side chain theory, to Pauling's instructive theory, or to Jerne's natural selection theory, all fail this test, and that is what, at the end, made Burnet's clonal selection theory—one lymphocyte, one anticipatory specificity—so convincing and attractive. Until some shared, but unusual molecular means of arranging for one specificity per lymphocyte is discovered, this phenomenon will not pass muster as proof of homology. Analogous arguments apply to the fact that both systems use only one of the two parental receptor loci (allelic exclusion—see Sect. 4.5.4), and that in both cases lymphocytes are stimulated to divide and clonally expand when activated.

So are there any features of the adaptive system in agnathans and in gnathostomes that can be fairly said to be homologous in an evolutionary sense? The answer is that there are least four major factors:

- They both possess lymphocytes
- They share particular transcription factors
- They both form specialised epithelium with the properties of a thymus
- They share the use of novel cytidine deaminases

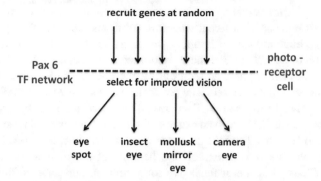

Fig. 4.12 Intercalary evolution of visual systems. Starting from a transcription factor (TF) network centred on Pax6, which specifies photoreceptor cells, genes may be recruited at random and selected on the basis of their ability to improve the visual system. In different species, different combinations are selected and these different combinations give rise to different eye forms [40]

4.16.2   Intercalary Evolution and "Deep Homology"

Nowhere have the questions of real and apparent homologies been investigated more extensively than in the evolution of visual systems. And nowhere have they been so deeply studied (see Appendix E). Numerous obviously different types of visual systems occur in metazoans, but all of them share the basic feature that a transcription factor cascade, headed by the master regulator Pax6, regulates their developmental program. This simple basic visual system was then subjected to improvement by the recruitment of genes more or less at random. Those genes that helped were retained; those that did not help were dropped. Since in different species the selective forces operating on the visual system are different, different genes have been selected in different species, and have given rise to different visual systems (Fig. 4.12).

Structurally and developmentally the various visual systems are all very different, but at a deeper level they are all homologous for all depend on the Pax6 transcription factor cascade, which is inherited by descent throughout phylogeny.

4.16.3   The Adaptive Niche

The Pax6 story tells us one more thing, and that is that there is no such thing as a "perfect" eye. Each visual system has been selected to fit the needs of very different animals in very different environments. Each species ends up with the sort of eye that gives it the maximal benefit for the minimal input of resources. The same sort of species-specific selective niches will also apply to immune defence, and they will have shaped the evolution of adaptive immunity. So what are the characteristics of the particular selective niche within which adaptive immunity evolved? Perhaps the central limiting factor is that a protein-based, anticipatory adaptive immune system requires a vast number of different receptors. With just one receptor specificity per lymphocyte, a huge number of lymphocytes must be constantly produced, and the majority of them do not survive the developmental process. Most of those that do will never meet their cognate antigen during the course of their short lives in the periphery, and so will die unused. The production and subsequent loss of all of these cells is metabolically expensive. This sort of adaptive immune system would be unlikely to help a small, short-lived organism with few immune cells—such as a fruit fly or a nematode worm. To make the investment worthwhile, the animal has to have features that will allow it to exploit the benefits of adaptive immunity so as to increase its fitness. For this it must live long enough, and have sufficient rapidly turned over immune cells, to be able to sample the huge repertoire during the course of its life. Somewhere at the start of vertebrate evolution the increasing "generation gap" between rapidly dividing pathogens on the one hand and ever-larger animals with longer generation times on the other, produced a favourable selective niche within which adaptive immunity made sense. After that the evolution of adaptive immune systems—while not inevitable—should not be considered to be a great surprise.

4.16.4 Haematopoiesis and the Origin of Lymphocytes

We all know—more or less—what a "lymphocyte" is. In vertebrates they are immune cells produced by the mesoderm-derived haematopoietic system, some, but not all of which, express somatically formed receptors of the adaptive immune system. Where did lymphocytes come from? In *Drosophila melanogaster* haematopoietic stem cells give rise to three innate immune cell types all of which are found in the haemolymph. The crystal cells and lamellocytes are specialised mobile cells, which encapsulate intruders like fungi or the eggs of parasitic wasps. This activity is somewhat reminiscent of the formation of granulomas in humans and mice, though these crystal cells and lamellocytes have no clear homologs in vertebrate systems. The third cell type—the plasmatocytes—are mobile phagocytic cells that remove cell debris and bacterial pathogens, and which in this sense are comparable to mammalian granulocytes or monocytes. These invertebrate lineages are considered homologous to vertebrate haematopoietic lineages, since they share with them the use of certain transcription factor cascades. However, in invertebrates there is nothing to compare to the range of blood cells produced by the haematopoietic stem cells in vertebrates. Though some invertebrates do have oxygen exchange molecules, either free in their body fluid or packed into the so-called plasmatocytes, none have an erythroid lineage that produces erythrocytes and there is nothing in invertebrate that can be compared in its developmental origin to the vertebrate lymphocyte lineages [41]. Somewhere down at the start of the vertebrate line there was a veritable explosion in the number of different cell lineages produced by haematopoietic stem cells.

When a stem cell lineage is extended so as to produce new cell types, as happened in the haematopoietic lineage as vertebrates first evolved, then each new cell type requires a network of transcription factors to organise the gene expression pattern, which will define its identity. These transcription factors will act to switch on the genes now appropriate for that new cell's activities, and to switch off genes that are inappropriate. Which transcription factors would be best for the new lineage? The answer is that it really doesn't matter—just as there is nothing about Pax6 that predestines it to be the master regulator of visual systems, so any transcription factor could in principle be recruited to regulate the development and the life of a new haematopoietic lineage. However, once a network of transcription factors has been chosen, and their short DNA recognition sequences inserted close to the start of the genes they should regulate, then it is very hard indeed to change the network. The target genes can be readily swapped by gaining or losing the short recognition sequences, but the transcription factors, once chosen, are fixed. A transcription factor network thus has an evolutionary stability that makes it a particularly powerful means of tracing homology, for there are so many transcription factors available that in cases of convergent evolution the probability that the same ones will be chosen by chance, is vanishingly low (see Appendix E).

We have no information as to how lymphocytes evolved, but by any reasonable definition of the term, cells which can be described as "lymphocytes" are present both in agnathans and in gnathostomes. Are these cells homologs, i.e. the product of one "invention" or are they two independent "inventions" that converged to do more or less the same job? Just as the Pax6 transcription factor demonstrates a deep homology between all bilaterian visual systems, one sees evidence of shared transcription factor networks in gnathostome and agnathan lymphocyte lineages. Perhaps the best evidence currently available comes from the transcription factor Pax5.

Pax5 is used throughout metazoan phylogeny in the specification of the nervous system, and it is also expressed in mice during spermatogenesis [42]. However, the extension of the haematopoietic lineages at the start of vertebrate evolution required that each of these new cell types be given a transcription factor network that would uniquely define its identity, and in the case of the B-lymphocytes in gnathostomes, identity is defined by a set of transcription factors that includes Pax5. No other cell in the haematopoietic lineage expresses Pax5, and a B-cell that loses its ability to express Pax5 ceases to be a B-cell. Thus, for example, Pax5 is switched off when a human or mouse B-cell turns into an antibody secreting plasma cell. In the agnathan lamprey there are three lineages of lymphocytes—the VLRA, B and C cells. Of these the VLRB cell is the only one, which, like the gnathostome B-cell, once activated releases its receptor in soluble form. It is also the only VLR cell to express Pax5, and this alone is a powerful argument that VLRB cells in the agnathans are homologous to B-lymphocytes in gnathostomes. In a similar vein VLRA and VLRC cells express transcription factors whose mammalian homologs are associated with $\alpha\beta$ T-cells and $\gamma\delta$ T-cells respectively [43]. By the criterion of transcription factor expression VLRA and VLRC cells are homologs to the gnathostome T-cell lineages, while the VLRB cells are homologs of gnathostome B-cells.

4.16.5   Thymus and "Thymoid"

The similarities between the agnathan and gnathostome systems do not end there. Gnathostome T lymphocytes develop in the thymus, an organ whose development is critically dependent on the expression of the transcription factor FOXN1. In the mouse this transcription factor is responsible for controlling the expression of a number of cytokines and cell surface molecules, including Scf, Cxcl12 and DLL4, and their expression is required to permit T-cell precursors to associate with the thymic micro-environment [44]. In the agnathan lamprey a region of the epithelium close to the gill tips expresses the agnathan homologs of both FOXN1 and DLL4. It is in this "thymoid" that the T-cell-like VLRA and VLRC cells rearrange their receptors. VLRB cells do not rearrange their receptors in this thymus homolog, but rather in a gut-associated tissue known as the typhlosole. The presence of a specialised FOXN1 and DLL4 expressing thymic-type lymphoid organ within which the T-cell-like lymphocyte lineages are formed, is a second strongly homologous feature shared by agnathans and gnathostomes, which must have been inherited from their last common ancestor.

4.16.6   Evolution of AID-Like Cytidine Deaminase Functions in Immunity

Early in vertebrate evolution there appeared a new family of cytidine deaminases. Members of this "Activation-Induced Deaminase" family are present both in agnathans and in gnathostomes, and so must have been present in their most recent common ancestor. These deaminase enzymes do not themselves cut DNA, but by converting dC in particular, defined DNA sequences into dU, they alert the cell's extensive array of DNA repair systems (Fig. 4.10). Depending on which of these mechanisms is recruited to deal with the dU, the result may be the induction of local point mutations, or of gene conversion, or of a larger chromosomal rearrangement [45].

In the agnathan lamprey, gene conversion forms the adaptive immune receptor repertoire, and the key players involved are AID-like cytidine deaminases (Fig. 4.2). In gnathostomes, in contrast, the means of forming the repertoire is mechanistically quite different. It is based not on gene conversion but rather on recombination catalysed by RAG (Figs. 4.5 and 4.6). This RAG recombination of adaptive immune receptor genes is restricted to gnathostomes, and hence was unlikely to have been present in the most recent common ancestor of agnathans and gnathostomes.

Since gene conversion and recombination are mechanistically quite different processes, one might expect that an adaptive immune system would be organised around one or the other. This would require that the switch from gene conversion in agnathans to recombination in gnathostomes was an abrupt saltation—one of those sudden transitions that do not happen often in evolution. But was it really that abrupt? There are a couple of observations that tend to make this transition seem rather less like a sudden switch. The first is that it appears that at the start of the evolution of gnathostomes AID and RAG may have worked together to form the primary repertoire. The reason for thinking this is that in the nurse shark—a basal gnathostome—the formation of the primary T-cell repertoire seems to involve both enzymes working in concert [46]. Furthermore, even in more highly evolved forms it turns out that AID's ability to generate the primary repertoire is a skill that has not been entirely forgotten. This can be seen in species like the chicken, in which all but one of the germline V-genes have accumulated mutations that converted them into pseudo-genes. How then does the chicken manage to form a diverse repertoire? The answer is that the one remaining functional germline V-gene is rearranged by RAG to form the initial receptor gene, and the diverse repertoire is then formed by AID-mediated gene conversion, which copies information into it from the flanking pseudo-genes. Thus, in species, which have failed to look after their germline V-genes properly, gene conversion can still be reactivated to save their adaptive immune system.

In this scenario AID-like deaminases were the key to immune receptor rearrangement in basal agnathans. Later, with the evolution of some primitive version of RAG-based recombination mechanism, AID and RAG worked in concert to generate the primary repertoire. Later still, the rearrangement process became dependent on RAG recombination alone. However, the RAG recombination process is intolerant of stop or frameshift mutations in the germline V-genes, and if these occur, the species quickly finds itself between a rock and a hard place. It then faces the choice of either trying to survive without adaptive immunity or reactivating repertoire formation by gene conversion. So far the latter choice appears to have been the only one that offered a chance of survival. As always, evolution rarely throws away a good idea, and though AID is no longer the primary generator of diversity in gnathostome adaptive immunity, it was not lost, but rather reassigned to novel roles—roles like the initiation of class switch recombination and of somatic hypermutation in germinal centre B-cells, which have no counterpart in the older agnathan system.

4.16.7   The Last Common Ancestor of Agnathans and Gnathostomes

Any attempt to reconstruct the evolution of adaptive immunity is bound to be highly speculative. Nevertheless, the molecular homologies between agnathans and gnathostomes allow us to make a reasonable guess as to the features that would have been required in the most recent common ancestor. The first of these is that the ancestor crossed the borderline of a selective niche, so that natural selection's cost–benefit analysis now made adaptive immunity a realistic option. The most recent common ancestor's haematopoietic stem cells generated a whole collection of new cell lineages, not found in basal invertebrates. One of these lineages led to the generation of three new cell types that were the precursors of the VLRB, VLRA and VLRC lymphocytes in agnathan and of the B-cells, CD4$^+$αβ T-cells, CD8$^+$αβ T-cells and γδ T-cells in gnathostomes. These precursor cell lineages in the most recent common ancestor were almost certainly mobile cells involved in immune defence. As such they would show another property crucial to mounting a concerted adaptive immune response—the ability to communicate with each other. This is achieved in gnathostomes either by direct cell-to-cell interactions or by the release by one cell of cytokines, which can be detected by the appropriate receptors present on other cells. IL-8, IL-16 and IL-17, along with their receptors, all play important roles in co-ordinating immune responses in gnathostomes and, though their functions have not yet been investigated in agnathans, it is clear that all of these cytokines and receptors are expressed by lamprey lymphocytes [43].

Just as in the visual systems, where a Pax6–photoreceptor cell axis (Fig. 4.12) can be expanded by random recruitment of genes, followed by selection for their ability to improve vision, so in adaptive immunity an axis defined by circulating immune cells on the one side, and a controllable mutagen-like cytidine deaminase on the other can then recruit in gene products that will provide for improved defence and hence increased fitness (Fig. 4.13).

One of the genes recruited must, of course, have coded for a cell surface receptor whose structure was such that it could be mutated so as to interact with a broad range of targets. One that "worked", in this sense, would be retained as the pathogen-binding receptor of the adaptive system. In the agnathan system we know nothing about the cell surface molecule, which gave rise to the receptors except that its extracellular domain was an LRR structure reminiscent of the TLR molecules of innate immunity. There are dozens of putative candidates in the genomes of modern cephalochordates or tunicates. In the gnathostome system, we need the Transib-invaded IgSF V-domain linked to an IgSF C1 domain. Both of these domains are found only in gnathostomes and so their route of evolution from invertebrate precursors remains speculative [47].

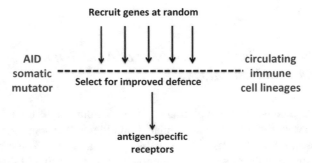

Fig. 4.13 Presumptive intercalary evolution of adaptive immunity in agnathans. Circulating immune cells expressing an AID-like cytidine deaminase mutator activity as somatic mutator, recruit genes at random. A recruited gene that can give rise to a family of receptors will improve fitness and be positively selected

One final point concerns the selection of the somatic mutators—cytidine deaminase in agnathans, and RAG in gnathostomes. Is it possible that RAG-based receptor rearrangement arose as an innate system backup to a cytidine deaminase-based adaptive system, like that in the lamprey? The reason for considering this is that there is an interesting anomaly in the earliest jawed vertebrates to which we have access. This involves an oddity in cartilaginous fish, in which the polypeptide chains of adaptive immune receptors are encoded by V, D and J segment that are rearranged by RAG recombinase. Unlike the situation in more advanced vertebrates, these gene segments are not organised into a single cluster within which recombination takes place. Instead there are many "mini clusters", each of which consists of just a few V, D and J segments. In many of these clusters the segments are partially recombined in the germline to D-J, V-D or even to fully recombined V-D-J elements. This tells us that in the ancestors of these fish, RAG recombinase must have been expressed in the germline, and raises the possibility that RAG recombination arose as a strategy to generate new, germline encoded, innate immune system receptors [48]. Perhaps RAG-based adaptive immunity in gnathostomes arose as a result of a molecular misunderstanding, when RAG, instead of being expressed in the germline, was suddenly expressed instead in lymphocyte precursors. A similar scenario might explain the initial selection of cytidine deaminase in agnathans.

Sadly, however, the details of the transition from agnathan to gnathostome adaptive immunity have been largely lost with the extinction of all agnathans, except for the basal groups represented by the lampreys and hagfish.

References

1. Wang E, Hunter CP (2017) SID-1 functions in multiple roles to support parental RNAi in *Caenorhabditis elegans*. Genetics 207(2):547–557
2. Tassetto M, Kunitomi M, Andino R (2017) Circulating immune cells mediate a systemic RNAi-based adaptive antiviral response in Drosophila. Cell 169(2):314–325 e13
3. Benitez AA et al (2015) Engineered mammalian RNAi can elicit antiviral protection that negates the requirement for the interferon response. Cell Rep 13(7):1456–1466
4. Schoggins JW et al (2011) A diverse range of gene products are effectors of the type I interferon antiviral response. Nature 472(7344):481–485
5. Adema CM (2015) Fibrinogen-related proteins (FREPs) in mollusks. Results Probl Cell Differ 57:111–129
6. Pancer Z et al (2004) Somatic diversification of variable lymphocyte receptors in the agnathan sea lamprey. Nature 430(6996):174–180
7. Holland SJ et al (2018) Expansions, diversification, and interindividual copy number variations of AID/APOBEC family cytidine deaminase genes in lampreys. Proc Natl Acad Sci U S A 115 (14):E3211–E3220
8. Han BW et al (2008) Antigen recognition by variable lymphocyte receptors. Science 321 (5897):1834–1837
9. Kapitonov VV, Jurka J (2003) Molecular paleontology of transposable elements in the Drosophila melanogaster genome. Proc Natl Acad Sci U S A 100(11):6569–6574
10. Kapitonov VV, Koonin EV (2015) Evolution of the RAG1-RAG2 locus: both proteins came from the same transposon. Biol Direct 10:20

11. Kapitonov VV, Jurka J (2005) RAG1 core and V(D)J recombination signal sequences were derived from transib transposons. PLoS Biol 3(6):e181
12. Watson CT, Breden F (2012) The immunoglobulin heavy chain locus: genetic variation, missing data, and implications for human disease. Genes Immun 13(5):363–373
13. Barclay AN (1999) Ig-like domains: evolution from simple interaction molecules to sophisticated antigen recognition. Proc Natl Acad Sci U S A 96(26):14672–14674
14. Wardemann H et al (2003) Predominant autoantibody production by early human B cell precursors. Science 301(5638):1374–1377
15. Rowland SL et al (2010) BAFF receptor signaling aids the differentiation of immature B cells into transitional B cells following tonic BCR signaling. J Immunol 185(8):4570–4581
16. Moore T, Haig D (1991) Genomic imprinting in mammalian development: a parental tug-of-war. Trends Genet 7(2):45–49
17. Lomvardas S et al (2006) Interchromosomal interactions and olfactory receptor choice. Cell 126 (2):403–413
18. Flajnik MF, Kasahara M (2010) Origin and evolution of the adaptive immune system: genetic events and selective pressures. Nat Rev Genet 11(1):47–59
19. Suurvali J et al (2014) The proto-MHC of placozoans, a region specialized in cellular stress and ubiquitination/proteasome pathways. J Immunol 193(6):2891–2901
20. Danchin EG, Pontarotti P (2004) Towards the reconstruction of the bilaterian ancestral pre-MHC region. Trends Genet 20(12):587–591
21. Flajnik MF, Kasahara M (2001) Comparative genomics of the MHC: glimpses into the evolution of the adaptive immune system. Immunity 15(3):351–362
22. Kaufman J (2018) Unfinished business: evolution of the MHC and the adaptive immune system of jawed vertebrates. Annu Rev Immunol 36:383–409
23. Nathan JA et al (2013) Immuno- and constitutive proteasomes do not differ in their abilities to degrade ubiquitinated proteins. Cell 152(5):1184–1194
24. Eisen HN et al (2012) Promiscuous binding of extracellular peptides to cell surface class I MHC protein. Proc Natl Acad Sci U S A 109(12):4580–4585
25. Klein L et al (2014) Positive and negative selection of the T cell repertoire: what thymocytes see (and don't see). Nat Rev Immunol 14(6):377–391
26. Ye P, Kirschner DE (2002) Reevaluation of T cell receptor excision circles as a measure of human recent thymic emigrants. J Immunol 168(10):4968–4979
27. Venkatesh B et al (2014) Elephant shark genome provides unique insights into gnathostome evolution. Nature 505(7482):174–179
28. Wilson M et al (1992) What limits affinity maturation of antibodies in Xenopus--the rate of somatic mutation or the ability to select mutants? EMBO J 11(12):4337–4347
29. Hwang JK, Alt FW, Yeap LS (2015) Related mechanisms of antibody somatic hypermutation and class switch recombination. Microbiol Spectr 3(1):MDNA3-0037-2014
30. Berek C, Milstein C (1987) Mutation drift and repertoire shift in the maturation of the immune response. Immunol Rev 96:23–41
31. Allen CD, Okada T, Cyster JG (2007) Germinal-center organization and cellular dynamics. Immunity 27(2):190–202
32. Vinuesa CG et al (2016) Follicular helper T cells. Annu Rev Immunol 34:335–368
33. Crotty S (2015) A brief history of T cell help to B cells. Nat Rev Immunol 15(3):185–189
34. Zhu C et al (2012) Origin of immunoglobulin isotype switching. Curr Biol 22(10):872–880
35. Manz RA, Thiel A, Radbruch A (1997) Lifetime of plasma cells in the bone marrow. Nature 388 (6638):133–134
36. Matsuda F et al (1998) The complete nucleotide sequence of the human immunoglobulin heavy chain variable region locus. J Exp Med 188(11):2151–2162
37. Star B et al (2011) The genome sequence of Atlantic cod reveals a unique immune system. Nature 477(7363):207–210
38. Tokunaga Y et al (2017) Comprehensive validation of T- and B-cell deficiency in rag1-null zebrafish: implication for the robust innate defense mechanisms of teleosts. Sci Rep 7(1):7536

39. Holland SJ et al (2014) Selection of the lamprey VLRC antigen receptor repertoire. Proc Natl Acad Sci U S A 111(41):14834–14839
40. Gehring WJ, Ikeo K (1999) Pax 6: mastering eye morphogenesis and eye evolution. Trends Genet 15(9):371–377
41. Hartenstein V (2006) Blood cells and blood cell development in the animal kingdom. Annu Rev Cell Dev Biol 22:677–712
42. Adams B et al (1992) Pax-5 encodes the transcription factor BSAP and is expressed in B lymphocytes, the developing CNS, and adult testis. Genes Dev 6(9):1589–1607
43. Hirano M et al (2013) Evolutionary implications of a third lymphocyte lineage in lampreys. Nature 501(7467):435–438
44. Calderon L, Boehm T (2012) Synergistic, context-dependent, and hierarchical functions of epithelial components in thymic microenvironments. Cell 149(1):159–172
45. Neuberger MS et al (2003) Immunity through DNA deamination. Trends Biochem Sci 28 (6):305–312
46. Ott JA et al (2018) Somatic hypermutation of T cell receptor alpha chain contributes to selection in nurse shark thymus. elife 7
47. Du Pasquier L (2004) Speculations on the origin of the vertebrate immune system. Immunol Lett 92(1-2):3–9
48. Lee SS et al (2000) Rearrangement of immunoglobulin genes in shark germ cells. J Exp Med 191(10):1637–1648

Further Reading

Boehm T, Hirano M, Holland SJ, Das S, Schorpp M, Cooper MD (2018) Evolution of alternative adaptive immune systems in vertebrates. Annu Rev Immunol 36:19–42
Flajnik MF, Kasahara M (2010) Origin and evolution of the adaptive immune system: genetic events and selective pressures. Nat Rev Genet 11:47–59
Janeway C (2017) Immunobiology, 9th edn. Taylor and Francis, New York
Kaufman J (2018) Unfinished business: evolution of the MHC and the adaptive immune system of jawed vertebrates. Annu Rev Immunol 36:383–409
Klein L et al (2014) Positive and negative selection of the T cell repertoire: what thymocytes see (and don't see). Nat Rev Immunol 14(6):377–391

Chapter 5
The Other Side of the Arms Race

Having read Chaps. 3 and 4, one might well be forgiven for assuming that pathogens should have long since ceased to be a problem. After all, immune receptor repertoires have been selected to "see" all sorts of different pathogens, and the information they generate can activate powerful effector mechanisms that should certainly be able to destroy all intruders. But we all know that this is not the case, and the reason is that, host–pathogen interactions are always an arms race (see Sect. 1.9), and so any effective defensive move made by the host merely provides the selection pressure that drives the evolution of new virulence strategies in pathogens. And indeed, random mutation and selection have provided pathogens with all sorts of ways to evade immune system receptors, to interfere with the subsequent signal transduction pathways, and to escape from terminal effector systems such as complement, granulocytes and $CD8^+$ T-killer cells. In its turn, every new pathogen virulence strategy drives the selection of new host resistance mechanisms. It is a never-ending struggle in which sometimes the pathogen, and then sometimes the host, manages to achieve a short-lived advantage.

In the struggle between pathogen and host both sides have clear priorities—both are out to maximise their fitness, and everything therefore revolves around a mathematical construction called the "Basic Reproductive Ratio" (R_0). This R_0 is defined as the number of secondary cases that derive from one infected individual. If the R_0 is less than 1.0, then the spread of the disease will be limited. However, if R_0 is greater than 1.0, then the number of infected hosts will increase, and the higher the value, the greater is the chance that an epidemic will result. The R_0 for any pathogen is dependent on a number of factors including the pathogen's mechanism of transfer, its ability to complete its life cycle in the host in the face of immune defence, and the host population structure.

A young and foolish pathogen may mount a virulent attack, so that the host suffers and soon dies. In this case the pathogen has to be able to quickly find a way of getting to a new victim if it is to achieve a R_0 value greater than 1.0. This is, for example, the route taken by *Yersinia pestis*, the causative agent of pneumonic plague. However, the structure of the host population may be such that the

© Springer Nature Switzerland AG 2019
R. Jack, L. Du Pasquier, *Evolutionary Concepts in Immunology*,
https://doi.org/10.1007/978-3-030-18667-8_5

pathogen's fitness is improved by moderating its attack. The life of the host is thus extended, and the pathogen now has a longer period in which to find its next victim. This sort of scenario was seen unfolding in real time in the 1950s in Australia, when highly virulent myxomatosis virus was deliberately released across the continent to control rabbit populations. The initial virus strain released was a "99.5% killer" with an average time from infection to death of 11 days. However, this level of virulence was not optimally suited to the structure of the rabbit population, and a mutant "90% killer" form that increased the infection-to-death period to 23 days soon appeared, and in the wild was naturally selected to replace the initial strain [1].

A reduction of pathogen virulence that extends the period available for transmission may lead to a *modus vivendi* in which a host population learns—one way or another—if not to tolerate the pathogen, then at least to coexist with it in a state of armed neutrality. *Staphylococcus aureus* is an example of this sort of thing. It lives permanently—but quietly—in the noses of one-third of humanity and is present intermittently in the noses of a further one-third. Normally it does no harm, but in conditions of physiological stress, or of reduced immune competence, *S. aureus* can turn into a monster that causes systemic sepsis and death [2].

5.1   Coming to Terms with Pathogens

For the host, one way to avoid the fitness loss associated with infection is to organise its life so as to reduce the risk of meeting the pathogen. The nematode worm *Caenorhabditis elegans* regards bacteria as either "edible" or "probably pathogenic". Quite how this distinction is made is unclear, but with the help of a TLR domain containing protein expressed in neurones, *Caenorhabditis elegans* manages to actively avoid areas on a plate where the pathogen *Serratia marcescens* is growing [3]. In a similar way, the larvae of *Drosophila melanogaster* avoid food that is contaminated with certain pathogens [4], and human beings also employ this sort of strategy by, for example, using condoms to avoid sexually transmitted diseases. However, it is not always possible to hold pathogens at bay by avoiding risky behaviour—a well-known example is *Mycobacterium tuberculosis* where the principle risk factor for acquiring the pathogen is breathing.

If the host cannot avoid the pathogen, then a second possibility may be to offer a deal that is in the best interests of both sides. An example of this is seen in the infection of the water flea *Daphnia magna* by the bacterial pathogen *Pasteuria ramosa*. The infection starts when the pathogen is eaten and attaches to the water flea's gut. The structures to which *Pasteuria* attaches are polymorphic, and different strains of the pathogen have evolved the ability to recognise one or another of these polymorphic forms. In this way the polymorphic resistance alleles of the water flea are matched to the polymorphic virulence alleles of the bacterium. When a particular clone of *Daphnia* expands, it will soon be attacked by the strain of pathogen whose virulence genes match it. This susceptible *Daphnia* population, together with its matching pathogen population, will then collapse as the water fleas are destroyed, and it will be replaced by a hitherto minor strain that is resistant to the current pathogen. Sooner or later this *Daphnia* population, in turn, will be found and attacked by a strain of the pathogen with suitable matching virulence alleles. This so-called negative frequency-dependent selection is a win-win situation for both sides, for none of the pathogen strains is able to infect all members of a heterogeneous *Daphnia* population, and equally, no individual *Daphnia* is resistant to all of the polymorphic strains of *Pasteuria*. The result is that *Daphnia magna* cannot be wiped out by *Pasteuria ramosa*, and *Pasteuria ramosa* will never run out of *Daphnia magna* to attack [5].

A third approach might be described as the "hired gun" strategy. In Chap. 4 of "On the Origin of Species" Darwin pointed out that, *"What natural selection cannot do, is to modify the structure of one species, without giving it any advantage, for the good of another species . . ."*. This is doubtless true, but what can indeed happen is that one species may enslave and ruthlessly exploit another to its own ends. For example, the protozoan *Toxoplama gondii* is a parasite that infects many warm-blooded animals, including rodents, but it can only complete its sexual reproductive phase in cats. It is therefore in Toxoplasmas's interest to end up in a cat, and this goal is supported by its ability to make the rodents it infects lose their fear of cats. A fearless rodent is one that is liable to soon become cat food, and in this way deliver *Toxoplasma* to the cat's intestinal epithelium. This way of doing things reaches its zenith in the case of the caterpillar of the Brazilian moth *Thyrinteina leucocerae*. A parasitoid wasp lays its eggs in the caterpillar of the moth. These eggs develop into larvae, which eventually cut a hole in the caterpillar's epidermis, exit and form pupae. At this stage these wasp pupae are defenceless, and hence liable to be eaten by any of a range of other insects. However, a couple of the wasp larvae are left inside the caterpillar and, in ways that are not understood, they take over the caterpillar's mind, turning it into a zombie guard. The caterpillar stops feeding, and instead stands over the pupae, violently swinging its head at any insect that approaches. In a field study, this reduced predation of the wasp pupae by around 50%. After a week, the wasps emerge from the pupae, and then the zombie dies [6].

Similar phenomena are also seen in immune defence strategies. In Sect. 1.11, we described the situation where Mimiviruses attack and kill amoebae. Some strains of amoebae have responded by recruiting the help of a Mavirus that parasitises the cytoplasmic site within which the Mimivirus assembles its progeny. The amoebae have integrated a silent copy of the Mavirus genome into their DNA, and there the integrated Mavirus sits quietly until such time as a Mimivirus attacks its host. The Mimivirus transcription factors activate this integrated silent copy of the mavirus, which then attacks the Mimivirus assembly site in the amoeba's cytosol. Many parasitic wasps that lay their eggs in the larvae of other insects use a similar sort of system. They protect their progeny within the host that they have parasitised by exploiting "domesticated" viruses integrated into their genomes to generate a variety of different virulence strategies involved in down-regulating the host's immune defence system [7].

5.2   Pathogen Strategies Directed Against the Innate Immune System

Pathogens work towards balancing virulence and transmissibility so as to maximise their fitness, and a major factor in this equation is, of course, their ability to counter the host's immune defences. Every pathogen has found some way of outwitting immunity, at least to some extent, and every one of them has found a different means of doing so. There are few broad general concepts as to how random mutation and selection help a pathogen to optimise its virulence strategy, though in vertebrates we can at least roughly divide pathogen strategies into those directed at evading innate immunity alone, and those directed at the much more ambitious goal of evading adaptive immunity as well. Most human pathogens probably try to do a bit of both, and they all use multiple virulence mechanisms to achieve their ends. This means that looking at pathogen virulence is rather a case of facing a vast list of details. Nevertheless, because these details end up being matters of life or death, it will be worth looking at a few examples.

5.2.1   Hiding from the Innate Receptor Repertoire

One strategy popular amongst pathogens is to find ways of becoming invisible to the limited number of microbe-detecting receptors available to innate immunity. Many pathogenic bacteria achieve this by the simple means of producing a capsule that covers the entire surface in a viscous coating. This thick gooey capsule is generally made of polysaccharide, though once again there is an inevitable exception—in this case it is *Bacillus anthracis,* whose capsule is made of polymerised glutamic acid. The capsule covers up microbial surface ligands that would otherwise be detected by innate receptors, and by doing so renders the encapsulated bacterium invisible. In addition, adaptive immunity is hampered because macrophages and dendritic cells have a hard time "getting their teeth into" the slimy surface, and so have difficulty phagocytosing them, to produce the peptide-MHC-complexes needed to initiate an adaptive response. Capsules are important virulence features, for their loss reduces or abolishes pathogenicity. Nevertheless natural selection never provides anyone with a truly free lunch. A capsule comes with a price that goes beyond the obvious investment of resources required to produce the slime, for it also cloaks the bacterial surface structures needed for interaction with host tissues, and thus limits the range of the pathogen's own virulence mechanisms.

Perhaps because a capsule comes with this price, some bacterial pathogens have evolved less drastic ways to achieve the same end. Immune defence is seldom an all-or-nothing affair, but rather a delicate balance between pathogen virulence on one side, and host resistance on the other. To tip this balance slightly, but decisively, in its favour a pathogen need not become entirely invisible. It may be sufficient for it to just momentarily move slightly out of focus. One example of this sort of thing involves *Yersinia pestis*—the causative agent of plague. *Yersinia pestis* caused three great pandemics: one in antiquity, one in the middle ages and one in the nineteenth century. Of these, the mediaeval pandemic is the best known. It first made itself apparent when in 1346 the Mongol army attacked the Genoese port city of Caffa on the Crimean Black Sea coast. The city was large, well defended by two concentric walls and could be supplied from the sea—so the siege dragged on. Then in the spring of 1347 the fighting tapered off as a sudden epidemic decimated the Golden Horde. The Mongols abandoned the siege, but before leaving they catapulted their dead comrades over the walls, bringing the sickness into the town. The citizens of Caffa died in their thousands, and from here the Black Death started its inexorable march through Europe [8].

Plague comes in two major forms—bubonic/septicaemic and pneumonic. These two forms have very different R_0 values. Bubonic plague is the commonest. It is spread by fleas that transfer it from infected rats. It is a serious illness, for if untreated many of the victims will die within 10 days. It is, however, not the stuff of which great pandemics are made, for its R_0 is generally limited. Pandemics arise when *Yersinia pestis* hits the jackpot by reaching the lungs of a plague victim. That is the point at which the disease turns into the pneumonic form, which then spreads rapidly by aerosol transmission. This is a vastly more efficient means of transmission than anything that fleas can manage, and so it was this pneumonic form that drove the three great pandemics. Nevertheless, at least initially, the commonest form is bubonic plague, and there *Yersinia pestis* has a problem. To achieve an R_0 value greater than 1.0 it must multiply rapidly, because it depends on fleas as its vector—and fleas do not drink large volumes of blood. Enormous numbers of bacteria must be present in the blood of a plague victim if a drinking flea is going to have a chance of picking them up. To achieve the necessary levels, *Yersinia pestis* must replicate at the highest possible rate. It has no time to lose, for it must complete the cycle of infection within 10 days or so, before the adaptive immune system kicks in. To win this race, *Yersinia pestis* comes with an impressive array of virulence factors [9]. One of the earliest acting ones centres on a structural change to the lipid-A part of the LPS molecule.

Lipid-A is the biologically active component of LPS that is recognised by the TLR-4 system (see Sect. 3.5.1). It is hydrophobic because it normally carries 6 fatty acid chains, and these are essential to its action. During biosynthesis of LPS the bacterium adds these six fatty acid chains in two phases—four in the first and two in a second. In *Yersinia pestis* the acyl transferase enzyme that adds the last two fatty acid chains is temperature sensitive. It works well enough in the flea, whose temperature is estimated to be around 26 °C, and there it produces the normal hexa-acyl LPS. However, once in the human host, where the temperature goes up to 37 °C, the enzyme is no longer active, and so the LPS formed only has four hydrophobic fatty acid chains. This "tetra-acyl Lipid-A" form is unable to stimulate human TLR-4. Using a mouse model of plague and a *Yersinia pestis* biovar associated with the mediaeval pandemic, it was shown that merely replacing this temperature sensitive acyl transferase with the non-temperature sensitive version from *E. coli* converted *Y. pestis* from a lethal pathogen into a harmless bacterium

[10]. It's worth bearing in mind here that TLR-4 is by no means the only innate receptor able to detect the presence of *Yersinia pestis*. However, failure of TLR-4 gives the pathogen a head start, perhaps by confusing the computations being carried out in the signal transduction network, and the result is that without the input from TLR-4 the other receptors are unable to mount an adequate, timely defence against the threat. On the other hand, if *Yersinia pestis* fails to dodge detection by TLR-4, then all of its other virulence mechanisms cannot save it. In this sense, tetra-acyl Lipid-A was the molecule that made the Black Death epidemic possible. Other pathogens, such as *Helicobacter pylori*, which causes stomach ulcers and is associated with the development of gastric cancer, have also learned to exploit the fact that the conversion of hexa-acyl Lipid-A to the tetra-acyl form renders their LPS invisible to the TLR-4 receptor system.

5.2.2 *Interfering with the Innate Response*

Blinding innate system receptors by capsule formation or dodging individual receptors with tricks like the use of tetra-acyl LPS is one means that pathogens employ to evade innate immunity. A second is by hacking the information processing system that converts detection of an incipient infection into a coordinated response. This system extends from the signal transduction pathways operating in each of the responding cells, to the cytokines and other molecules that permit these cells to share information with other cells. This information processing network is a prime target for pathogens, and there is probably not a single intracellular signal transduction pathway or intercellular information transfer system that is not manipulated by one pathogen or another.

Terminal effector systems have also been subverted by pathogens. Phagocytosis is a case in point. Though it is a major mechanism used for the destruction of microbial intruders, by no means every pathogen is interested in avoiding it, and some hijack the phagocytosis pathway to establish an intracellular niche for themselves. Once taken up by phagocytosis such pathogens must somehow escape fusion of the endosome with lysosomes. There is more than one way of doing this. Some, such as *Listeria monocytogenes*, dissolve the endosome membrane and escape into the cytosol. *Salmonella*, on the other hand, avoids death at the hands of lysosomes in a quite different way. It converts the endosome into a safe haven by exporting from it mediators, which block the cell's microtubule system needed to fuse endosomes to lysosomes [11]. In this way *Salmonella* establishes a replicative niche within the cell where it is safe not only from the lysosomes but also from all the innate immune receptors in the cytosol. The host, however, has a response to this: a protein in the endosome membrane—NRAMP-1—pumps divalent cations out of the endosome vacuole and so starves the bacterium of these essential growth factors. Pathogens like *Salmonella* respond by synthesising and secreting small molecules called siderophores that are able to bind divalent metal ions with exceedingly high affinity. The bacterium then uses a receptor to bind and internalise any siderophore molecules that have managed to scavenge a metal ion. One can see from all this that the endosome has been—and is—a major battleground, and bacterial pathogens like *Salmonella* must constantly struggle to keep their niche habitable, as the host struggles to make life within that niche as unpleasant as possible.

5.3   Examples of Pathogen Strategies to Avoid Adaptive Immunity

Pathogens that have evolved an infection cycle requiring their survival in a vertebrate host for more than 10–12 days must acquire the capacity to evade adaptive immunity. The adaptive system too relies on information processing within and between cells, and pathogens have evolved the means to interfere with these processes at many levels. Yet adaptive immunity is a much harder nut for a pathogen to crack, because it has a vastly greater number of different receptors than does innate immunity, and attempts to hide structures from adaptive immunity is made much more difficult because the receptors of αβ T-cells detect their targets indirectly in the form of peptides. The message from adaptive immunity to most pathogens is summed up in Joe Louis's slogan, "*You can run, but you can't hide*". Nevertheless pathogens, such as the Human Immunodeficiency Virus (HIV), the Human Cytomegalovirus (HCMV), *Mycobacterium tuberculosis* or *Trypanosoma brucei*, have indeed managed to devise means to hide, and so to survive in the face of adaptive immunity. The typical strategies that they use may be summed up under the headings "brute force", "playing dead" and "hiding in the future", and of course many pathogens use more than one of these simultaneously. Before looking at a few examples it is worth noting that mechanisms of pathogen evasion of adaptive immunity is an area that is notoriously difficult to study. Immunity is a complex system involving many different cell types that exchange information in organised lymphoid tissues, and these processes cannot be adequately modelled in in vitro experiments. Animal experiments, though more difficult to design and carry out, are much more informative, but they also have their limits, for immune systems in different species have been selected to cope with different environments, and hence with different spectra of pathogens. By no means all infections in humans can be precisely modelled in animal experiments.

5.3.1   Disrupting Adaptive Immunity's Detection Hardware

One strategy that is used by pathogens to establish a long-term infection is to directly attack the adaptive immune system. Perhaps the best known example of this sort of thing is provided by HIV, which goes straight for the jugular by infecting and destroying activated CD4$^+$ T-cells—the central controlling cells of adaptive

immunity. By doing so the virus ensures that in untreated patients the capacity of
adaptive immunity to fight the infection becomes progressively reduced.

HCMV too is able to outwit adaptive immunity, though it employs rather more
subtle approaches than does HIV. HCMV can't actually evade adaptive immunity,
but it can go a long way to slowing it down. Chief amongst the mechanisms that it
uses to do this are those that reduce the effectiveness of the CD8$^+$ T-killer cells that
will destroy virus-infected cells (Sect. 4.12.1). Killer T-cells detect MHC-Class I–
viral peptide complexes expressed on the surface of infected cells. These complexes
are formed when viral proteins present in productively infected cells are broken
down into peptides by the proteasome, transported by the "TAP" complex into the
endoplasmic reticulum where they bind to MHC-Class-I molecules and then are
transferred to the cell surface. HCMV interferes with these processes at a number of
points so that insufficient MHC-Class-I complexes are expressed on the infected
cells, making them essentially invisible to the CD8$^+$ killer T-cells. That would seem,
at first blush, to be a winning move on the pathogen's part, but many viruses do this
sort of thing, and so not surprisingly natural selection has provided immunity with an
innate system backup consisting of "Natural Killer" (NK) cells. NK cells are
lymphocytes that do not express T- or B-cell receptors, but instead they express a
set of innate system receptors that recognise MHC-Class-I molecules (see Sect.
3.6.6). They patrol through the body screening other cells for their ability to express
MHC-Class-I on the surface. If they meet cells that do not express MHC-Class-I,
then the bet is that these are virus-infected cells, and the NK cells will kill them. This
screen for "missing self" might seem to be a KO blow to the virus, but HCMV has
struck back by evolving a gene whose product is expressed on the surface of an
infected cell, and looks for all the world like an MHC-Class-I molecule. And so it
goes on, and on, and on.

5.3.2 Playing Dead

A pathogen trying to avoid the attentions of adaptive immunity may try to make
itself as unobtrusive as possible by establishing a latent infection in which—ideally—it would become invisible by hiding quietly in some cell type or structure. It is
certainly possible to go far down this road, but the more successful a pathogen is in
this endeavour—the more invisible it becomes—the less chance it has of organising
its transmission to a new host. Perhaps for this reason pathogens pursuing this
strategy tend to maintain two populations within the host; one hiding away in the
latent state while a second, active population faces the perils of adaptive immunity
and looks for a chance to move on. This may lead to a situation in which the interests
of these two populations begin to diverge a little. The active pathogen population
may enter an evolutionary dead-end, forgetting about new hosts and instead measuring its success in terms of its ability to infect new cells in the current host. The
latent population, in contrast, may be waiting for some sign, perhaps provided by the
active population, that immunity is a bit under the weather, so that it is now

opportune to wake up and seize the chance for transmission to a new host. To some extent one sees this divergence in the infection strategy followed by HIV. On transmission HIV infects $CD4^+$ T-cells, though not all such cells are treated equally. Some $CD4^+$ memory T-cells may be latently infected—they contain virus that "plays dead". However, the majority of infected cells are productively infected and hence continually making virus that will infect new cells. That is a problem for the pathogen, because for HIV infection is a highly mutagenic process.

When HIV enters a new cell it first converts its RNA genome into DNA by using a virus-encoded reverse transcriptase. At this stage of its life cycle it risks picking up lethal levels of mutation for its reverse transcriptase is error prone. Furthermore, though the host's APOBEC3G cytidine deaminase, which introduces mutations into the viral genome during reverse transcription (see Sect. 3.6.1), is targeted for destruction by the product of HIV's *vif* gene, this process is incomplete so that APOBEC3G's mutagenic effect, though reduced, is not abolished. Thus every time the virus infects a new cell it is being mutated and the mutants are then selected for their ability to survive and themselves then infect new cells. Optimisation for long-term survival within the current host is not necessarily going to produce the ideal characteristics for optimal transmission. For that, reactivation of the latent virus may be the better bet, and this may explain why the rate of HIV evolution by hypermutation within a host is more rapid than the rate of evolution of HIV sequences when followed at the epidemiological level [12].

Mycobacterium tuberculosis—the greatest bacterial killer of human beings bar none—is a second pathogen that also evades adaptive immunity by playing dead. An estimated one in three humans is infected, though around 90% of those infected do not suffer any ill effects—they are said to be latently infected. Why is this? *Mycobacterium tuberculosis* is normally controlled both by macrophages that phagocytose the pathogen, and by a T-cell response. The macrophages, it must be admitted, find the bacterium rather indigestible, so what they do is to build an organised structure called a granuloma. Granulomas are used by many organisms, both vertebrate and invertebrate, to enclose and encapsulate pathogens or inanimate particles. Different types of granuloma are formed in response to different challenges within one host species, and the response to a given challenge can be different in different host species [13].

The granulomas encapsulating *M. tuberculosis* in human lungs contain both macrophages and T-cells and may be surrounded by a fibrotic coat. The centre of the granuloma is practically anoxic and can be full of dead cells. The formation of these structures sounds like a splendid host scheme to control the pathogen—but the pathogen has evolved an astonishing array of mechanisms to counter this seemingly fool proof host response [14]. First, *M. tuberculosis* has a truly remarkable capacity to survive in a dormant state in the unfriendly environment of a granuloma. Second, even in this latent disease state, active bacteria—often referred to as "scouts"—are present outside of the granuloma. The latent infection thus represents a balance between the immune system, the active "scouts" and the reservoir of dormant bacteria in the granulomas. Any alteration in the balance of these factors, such as for example an HIV-induced reduction in the effectiveness of the immune system, can result in reactivation and disease. How this reactivation is effected is a matter of controversy, though the active scouts may secrete soluble "resuscitation promoting factors" that help reawaken dormant bacteria. This strategy makes *M. tuberculosis* as close to a perfect pathogen as you can get, and that is what makes it so difficult to counter.

Another human pathogen that also can play dead is HCMV. In economically developed countries it infects more than 40% of the population and in underdeveloped countries close to 100%. HCMV is specific for human beings, and because of the limited value of work with cell lines, and the lack of a suitable animal model, it is extremely difficult to study the infection cycle of this pathogen. It is, however, clear that the virus can infect many different cell types in a human being, and it is also clear that the result of the infection varies from cell type to cell type. In some cell types, such as haematopoietic progenitor cells, the virus establishes a latent infection and so "plays dead". In other cell types, however, the infection is productive in the sense that new virus particles are produced, and yet in a healthy individual, though virus may be secreted in the urine and saliva, there are no overt signs of infection [15]. The virus is thus able to spread, but it manages to do so while at the same time establishing a balance with the immune system that makes possible a life-long subclinical infection. If this balance between host and pathogen is disturbed by a failure of the immune system, for example in transplant patients or in those with AIDS, then the virus can escape control and cause life-threatening illness.

5.3.3   Hiding in the Immunological Future

A third strategy that is used by pathogens to evade adaptive immunity is to constantly change shape, so that by the time immunity has selected and expanded lymphocytes whose receptors will recognise it, it has turned into something else. HIV indulges in this by virtue of the hypermutation of its genome during reverse transcription, for this generates escape variants to virtually anything that immunity can come up with. However, this way of doing things reaches its highest form in the "antigenic variation" strategies employed by certain eukaryotic parasites.

African sleeping sickness is caused by infection with the eukaryotic pathogen *Trypanosoma brucei* that is transmitted by tsetse flies. This unicellular parasite swims around in the blood of an infected individual, propelling itself with the help of a flagellum. Since it doesn't hide inside cells or in granulomas, it is exposed to all of the armaments of innate and adaptive immunity. Despite this, it survives. How does the trypanosome manage this? The answer has to do with the nature of its surface layer, which has been selected to be resistant both to innate and to adaptive immunity. In the blood the trypanosome is covered with many copies of a "variable surface glycoprotein" (VSG). VSG molecules are long and thin, so that they physically screen everything else on the trypanosome surface from the immune system. All that immunity can see are the outer tips of the long thin VSG protein molecules, and these do not provide a good target for any of the innate system's collection of pentraxins, collectins, ficollins and so on, or for natural antibodies, which might otherwise lead to complement activation and destruction of the parasite. Of course nothing in life is perfect, and so now and then an immune receptor will indeed bind to the surface, but the trypanosome has evolved a means of escaping from such attacks. It achieves this by recycling the surface molecules at a very rapid rate. As the trypanosome, propelled by its flagellum, moves through the blood, its surface VSG molecules are constantly being swept backwards towards the flagellum where they are taken into the cell and then recycled to the surface. It takes just over 12 min to turn over the entire surface VSG pool in this way. Furthermore, a protein binding to a VSG molecule acts like a sail, and so the complex is swept by the flow of liquid across the trypanosome's surface, back towards the flagellum at a much faster rate. Because of this a VSG molecule with an innate receptor or natural antibody bound to it is cleared from the surface within an estimated 120 s.

The trypanosome thus has little to fear from innate immunity, but adaptive immunity ought to be a different matter. The trypanosome would have a hard time making its VSG-covered surface invisible to adaptive immunity, and it doesn't even try to do this. In fact the VSG molecules are splendid immunogens and the host generates a strong antibody response against them. These antibodies saturate the trypanosome surface and cannot be swept away fast enough to prevent certain death for the parasite. The majority of the trypanosomes may be killed, but a small fraction of the population is constantly generating variant VSG structures against which the antibodies are useless. This population of trypanosomes with a new VSG structure has "escaped" into the immunological future. The "escaper" population rapidly expands, and the immune system starts once again to make antibody against the new VSG. No sooner are they ready than a small fraction of the population once again starts expressing yet another new VSG structure. To be successful at this the trypanosome has to be able to constantly come up with novel VSG structures. It achieves this by starting off from a collection of around 2000 genes coding for VSGs. However, the number of different VSG molecules that can be generated is very much larger than this number would suggest because the information in these 2000 genes (many of which are pseudo-genes) can be mixed and matched by gene conversion to produce hybrid VSGs [16]. In essence, this is a mirror image of the genetic trick used by agnathans to form their adaptive immune defence system (Sect. 4.3)—but this time it is giving rise to a virulence mechanism that overwhelms adaptive immunity.

In terms of host–pathogen interactions it is worth comparing this situation, where the host's resistance factors and the pathogen's virulence factors are both being generated by somatic hypermutation, to that of *Daphnia* and its bacterial pathogen *Pasteuria,* where the interaction is based on sets of polymorphic virulence and resistance factors that are "in place" within the respective populations. For the *Daphnia–Pasteuria* interaction neither side has an inevitable temporal advantage, and the resulting negative frequency-dependent selection works in the interests of both. However, replacing competing sets of polymorphic genes with ligands and receptors generated by somatic hypermutation leads to a very different result, for here the trypanosome has the enormous advantage that it is always given the first move. The host then has to respond—but it responds with a delay that is determined by the time needed to select and expand lymphocytes expressing the appropriate adaptive immune system receptors. It is this delay—inherent in the way adaptive immunity generates its receptors—that gives the trypanosome its trump and makes sleeping sickness inevitably fatal unless medication is available.

References

1. Burnet FM (1959) The clonal selection theory of acquired immunity. Cambridge University Press, Cambridge
2. Darisipudi MN et al (2018) Messing with the Sentinels-The interaction of *Staphylococcus aureus* with dendritic cells. Microorganisms 6(3)
3. Pradel E et al (2007) Detection and avoidance of a natural product from the pathogenic bacterium *Serratia marcescens* by *Caenorhabditis elegans*. Proc Natl Acad Sci U S A 104 (7):2295–2300
4. Babin A et al (2014) Fruit flies learn to avoid odours associated with virulent infection. Biol Lett 10(3):20140048
5. Bento G et al (2017) The genetic basis of resistance and matching-allele interactions of a host-parasite system: the *Daphnia magna-Pasteuria ramosa* model. PLoS Genet 13(2):e1006596
6. Grosman AH et al (2008) Parasitoid increases survival of its pupae by inducing hosts to fight predators. PLoS One 3(6):e2276
7. Pichon A et al (2015) Recurrent DNA virus domestication leading to different parasite virulence strategies. Sci Adv 1(10):e1501150
8. Wheelis M (2002) Biological warfare at the 1346 siege of Caffa. Emerg Infect Dis 8(9):971–975
9. Zhou D, Yang R (2009) Molecular Darwinian evolution of virulence in *Yersinia pestis*. Infect Immun 77(6):2242–2250
10. Montminy SW et al (2006) Virulence factors of *Yersinia pestis* are overcome by a strong lipopolysaccharide response. Nat Immunol 7(10):1066–1073
11. D'Costa VM et al (2015) Salmonella disrupts host endocytic trafficking by SopD2-mediated inhibition of Rab7. Cell Rep 12(9):1508–1518
12. Fraser C et al (2014) Virulence and pathogenesis of HIV-1 infection: an evolutionary perspective. Science 343(6177):1243727
13. Pagan AJ, Ramakrishnan L (2018) The formation and function of granulomas. Annu Rev Immunol 36:639–665
14. Ehlers S, Schaible UE (2012) The granuloma in tuberculosis: dynamics of a host-pathogen collusion. Front Immunol 3:411
15. Goodrum F (2016) Human cytomegalovirus latency: approaching the Gordian knot. Annu Rev Virol 3(1):333–357
16. Hovel-Miner G et al (2015) A host-pathogen interaction reduced to first principles: antigenic variation in *T. brucei*. Results Probl Cell Differ 57:23–46

Further Reading

Darwin C (1859) On the origin of species. John Murray, London
Gengenbacher M, Kaufmann SHE (2012) *Mycobacterium tuberculosis*: success through dormancy. FEMS Microbiol Rev 36:514–532
Goodrum F (2016) Human cytomegalovirus latency: approaching the Gordian knot. Annu Rev Virol 3(1):333–357
Hsu E, Du Pasquier L (2015) Pathogen-host interactions: antigenic variation v. somatic adaptations. Results and problems in cell differentiation, vol 57. Springer, Switzerland

Chapter 6
Postface

Here, at the end, it's appropriate to go back to the beginning—to Dobzhansky's premise that "nothing in biology makes sense except in the light of evolution". Is this really true for immunity? We think it is. The value of the light that an evolutionary approach can shed is seen in the regularity with which insights gained from "simpler" systems have led to advances in our understanding of how the mammalian immune system functions. Yet, the light of evolution is not a spotlight, but rather a diffuse source that leaves some aspects of the subject making sense, while others remain under-illuminated. The dark areas are caused in part by the extinctions during phylogeny, which leave the answers to many specific questions, such as the origin of MHC molecules, or the switch from agnathan to gnathostome adaptive immunity, buried in the mists of evolutionary time. In part, they are also due to the fact that invertebrate systems are often felt to be of secondary interest, and hence they remain largely unexplored.

However, in the brightly lit areas, the illumination does help make sense of the weird structure of immune defence systems. The evolution of immunity was largely driven by the arms race between pathogens and hosts, in which every new pathogen virulence mechanism fuelled the selection of new host virulence strategies, which in turn drove the selection of new pathogen virulence mechanisms, and so on, and on, and on. Thus the evolution of immunity is not the account of an orderly progression, but rather a tale of how pathogen and host both stagger from one life-threatening emergency to the next. And when one is fighting for one's life, one does not sit down to design the best and most elegant solution, one simply grabs whatever is lying around, and tries to see if it can be used as a weapon. In this sense, evolution's central trademark is to borrow old genes and then refurbish them to solve new problems.

A borrowed gene may be tweaked by mutation, or new functions may be cobbled together by exon shuffling from bits and pieces of several different old genes. Borrowing and recombination are the central resources that provide a genome under attack with the possibility of coming up with new moves to counter the pathogen. Sometimes these solutions may seem to us to be very strange indeed:

© Springer Nature Switzerland AG 2019
R. Jack, L. Du Pasquier, *Evolutionary Concepts in Immunology*,
https://doi.org/10.1007/978-3-030-18667-8_6

An engineer or systems analyst, who suggested the astonishingly wasteful RAG recombination process as a solution to the problem of pathogen attack, would be fired on the spot. But evolution is not interested in the elegance of a system; the only criterion that evolution applies is, "Does it work?" And sometimes very odd things do indeed work. RAG recombination "works" in the sense that the fitness costs it brings with it are more than matched by the fitness gains that an adaptive immune system confers.

The evolution of immunity follows the principle of "permanent revolution". Nothing stays quite the same. The molecules may be retained, but the functions can be changed. For example, the anti-viral cGAS-STING pathway has been preserved as an antiviral defence system across a huge span of phylogeny, but the mechanistic basis for its function has changed dramatically from invertebrates to vertebrates. Likewise, molecules of the Toll family play important roles in innate immunity, both in invertebrates and in vertebrates, and yet the role of Toll in Drosophila is very different from that of TLRs in vertebrates.

At the same time, this revolutionary principle is moderated by a conservative principle of redundancy. Thus adaptive RNAi-based anti-viral defence in invertebrates is backed up by the innate cGAS-STING pathway, which, in a rather different form, also backs up adaptive $CD8^+$ T-cell-based anti-viral defence in vertebrates. Recognising all this may help students to de-mystify their view of immunology, to see it less as a mass of details and rather more as one of the currently most fascinating and challenging aspects of biology.

Finally, and best of all, our current understanding of the subject is no more than a mere snapshot of the present, for immunity is constantly changing as it tracks the evolution of pathogens, and that will guarantee that the children and grandchildren of immunologists today will certainly find gainful employment, and have ever-newer fascinating problems to solve, in the future.

Berlin and Basel, January 2019

Appendices

Appendix A: A Simplified Classification of Metazoa

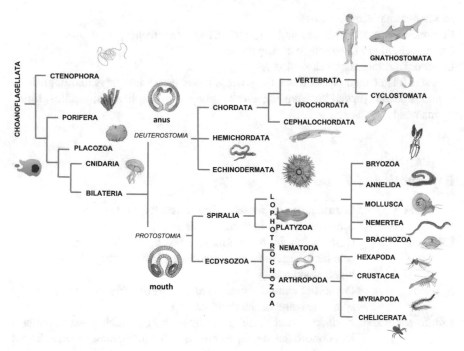

Fig. A.1 A simplified classification of Metazoa. Explanation of terms and positions in phylogeny of animals mentioned in the text

© Springer Nature Switzerland AG 2019
R. Jack, L. Du Pasquier, *Evolutionary Concepts in Immunology*,
https://doi.org/10.1007/978-3-030-18667-8

Coanoflagellates: *Monosiga brevicollis*

Ctenophora: comb jellies
Porifera: sponges, e.g. *Suberitis domuncula, Amphimedon queenslandica*

Placozoa: e.g. *Trichoplax adherens*
Cnidaria: jellyfish, sea anemone, e.g. *Nematostella vectensis*

Nematoda: round worms, e.g. *Caenorhabditis elegans*
Platyzoa: flat worms, e.g. *Macrostomum lignano*

Bryozoa: moss animals
Annelida: segmented worms
Mollusca: snails, squids, oysters
Brachiopoda: brachiopods
Nemertea: ribbon worms
Hexapoda: insects, e.g. *Drosophila melanogaster*
Crustacea: shrimps, crabs, barnacles
Myriapoda: millipedes
Chelicerates: spiders, ticks, scorpions

Hemichordata: acorn worms
Echinodermata: sea stars, sea urchins, e.g. *Strongylocentrotus purpuratus*
Cephalochordata: Lancelets, e.g. *Amphioxus*
Urochordata: tunicates, sea squirts
Agnatha (also called Cyclostomata): (jawless vertebrates), lampreys and hagfish
Gnathostomata: (jawed vertebrates), sharks, bony fish, amphibians, reptiles, birds,
    mammals

Explanation of Terms

Cnidaria   Gastrulation in these animals results in two germ layers—
                      ectoderm and endoderm.
Bilateria   Gastrulation in these animals results in three germ layers—
                      ectoderm, mesoderm, endoderm. Such "triploblasts" have
                      bilateral symmetry.
Spiralia   Phyla that exhibit a canonical pattern of early development, the
                      spiral cleavage of the early embryo.
Lophotrochozoa   A clade within Spiralia the members of which possess either a
                      lophophore (crest-shaped ciiature) or trochophore (wheel-shaped
                      ciiature) larvae.
Ecdysozoa   Include all animals that grow by ecdysis, i.e. by moulting their
                      exoskeleton.

All of the animals shown use forms of innate immunity for their defence. Adaptive immune systems, which use nucleic acids as sensors, have been studied in nematodes and arthropods. Adaptive immune systems that use proteins composed of Leucine rich Repeats as sensors have been studied in Agnathans. Gnathostomes are equipped with adaptive immune systems that utilise as sensors proteins composed of Immunoglobulin Superfamily Domains.

Appendix B: Immune Receptors and Their Common Domains

	Found in	Contains domains								Ligands, targets
Adaptive receptors										
BCR, TCR	G	IgSF								Any
VLR	Ag		LRR							
Innate receptors										
MHC-class I	G	IgSF								Peptides
MHC-class II	G	IgSF								Peptides
CD14	V		LRR							Lipopolysaccharide
TLRs	V		LRR							MAMPs, DAMPs
NLR	M		LRR	NBD	CARD					MAMPs, DAMPs
RIG-I	M				CARD	Hel				RNA
DICER	A					Hel				RNA
Lectins	M						Ctl			MAMPs
MBL, Collectin	M						Ctl	Col		MAMPs
Ficolin	M							Col	Fib	MAMPs
FREP	Mol, A	IgSF							Fib	Parasites
AID	V									CDD	DNA
APOBEC	Mam									CDD	DNA
Galectin	Metazoa	CRD								ß-galactosides
PGRPs	Most M	Type 2 amidase								Peptidoglycans
CRP	M	Pentraxin domain								MAMP
STING	F,M	CNBD								Cyclic di-nucleotides

Somatically formed adaptive and germ-line encoded innate immune receptors are shown together with the phylogenetic groups in which they occur (*Ag* agnathans, *A* arthropods, *B* bilateria, *F* flagellates, *G* gnathostomes, *M* metazoa, *Mam* mammals, *V* vertebrates). The principle domains that they contain and the ligands that they bind are indicated. Abbreviations: *AID* activation-induced deaminase, *APOBEC* apolipoprotein B mRNA-editing enzyme, catalytic polypeptide, *BCR* B-cell receptor, *CARD* caspase activation and recruitment domain, *CDD* cytidine deaminase domain, *CNBD* cyclic nucleotide-binding domain, *CRD* carbohydrate recognition domain, *CRP* c-reactive protein, *Ctl* C-type lectin domain, *Col* collagen domain, *DAMP* danger-associated molecular pattern, *Fib* Fibrinogen domain, *FREP* Fibrinogen-related proteins, *Hel* helicase super family 2, *IgSF* immunoglobulin super family, *LRR* leucine rich repeat, *MAMP* microbe associated molecular pattern, *NBD* nucleotide-binding domain, *NLR* NOD-like receptor, *PGRP* proteoglycan receptor protein, *TCR* T-cell receptor

Appendix C: cGAS, STING and Cyclic Dinucleotides

In bacteria, cyclic dinucleotides serve as second messengers and are required for the regulation of numerous essential cellular processes. Not surprisingly their synthesis and turnover are stringently controlled and there are dozens of proteins that regulate their half-lives. In Gram-negative bacteria, the most important cyclic dinucleotide is composed of two guanosine monophosphate (GMP) molecules linked by phosphodiester bonds between the 3′ hydroxyls and the 5′ phosphates to produce a "3,3 cyclic di-GMP" (Fig. C.1a). In Gram-positive bacteria a 3′,3′cyclic di-adenosine is also produced, while in the G⁻ human pathogen *Vibrio cholerae* there is a mixed cyclic dinucleotide, 3′,3′cyclic GMP-AMP, formed.

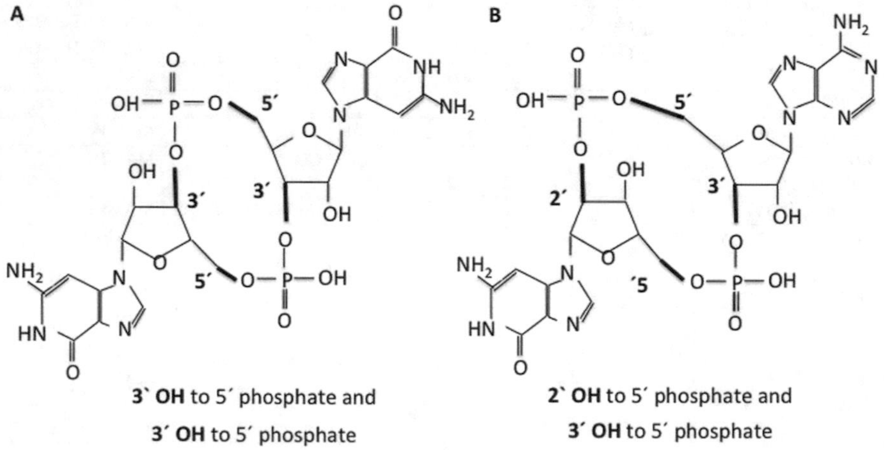

A

3ˋ OH to 5′ phosphate and
3′ OH to 5′ phosphate

B

2ˋ OH to 5′ phosphate and
3′ OH to 5′ phosphate

Fig. C.1 Structures of the cyclic di-purines (guanine and/or adenine). (**a**) Two purine moieties linked by phophodiester bonds between the 3′ hydroxyls and the 5′ phosphates. (**b**) Structure of the cyclic GAMP in which a guanine (left) is linked by a 2′ hydroxyl to the 5′ phosphate of an adenine, which is linked via its 3′ hydroxyl to the 5′ phosphate of the guanine

In eukaryotes, in contrast, cyclic dinucleotides are not nearly so important for the regulation of cell physiology. However, in the sea anemone *Nematostella vectensis* there is a cyclic GMP-AMP-synthase (cGAS), which generates a 3′,3′cyclic GMP-AMP (cGAMP). It is not known what activates this sea anemone cGAS, but the cGAMP it produces is bound by an anemone homolog of the vertebrate protein "Stimulator of Interferon Genes" (STING). This name, we must hasten to add is a bit of a misnomer in this context, for interferon genes, being a vertebrate "invention" are not present in the sea anemome. In *Drosophila* there is also a cGAS enzyme, but again we don't currently know what switches it on. *Drosophila* also has a STING homolog, which may bind cyclic dinucleotides derived from bacterial pathogens, after which it induces an innate immune response.

In vertebrates, however, the story changes. The cGAS enzyme has acquired a zinc ribbon domain, which enables it to bind DNA. Once DNA is bound the vertebrate cGAS produces a cGAMP, but it differs from the invertebrate product in that the 2′ hydroxyl of the GMP is linked to the 5′ phosphate of the AMP, while the 3′ hydroxyl of the AMP is linked to the 5′ position of the GMP (2′,3′ cyclic GAMP) (Fig. C.1b). This cyclic dinucleotide is bound by the vertebrateSTING which then induces the interferon response.

Appendix D: Immunoglobulin Superfamily Domain

The immunoglobulin superfamily domain is the name given to a protein fold first identified in immunoglobulins, but now known to be present in many eukaryotic proteins [1]. The IgSF structure consists of a series of short β strands connected by loops, as shown in the figure. There are even bacterial proteins which have this structure, though it is possible that these were acquired from eukaryotes by horizontal gene transfer [2]. There are, however, a number of bacterial proteins, which have no significant sequence homology to eukaryotic IgSF proteins, yet nevertheless have a similar three-dimensional structure.

The V, C2 and I forms of this domain are present throughout phylogeny. The CI domain is restricted to gnathostomes. In the case of the gnathostome adaptive immune receptors, the antigen-binding domains are made up of the loops joining strands B and C, G and F and C′ and C″.

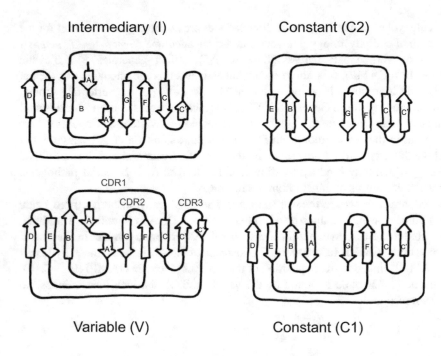

Intermediary (I)   Constant (C2)

Variable (V)   Constant (C1)

IgSF domains

Appendix E: Intercalary Evolution

E.1 What Visual Systems Tell Us About Homology and Analogy

The ability to see confers a huge selective advantage, and thus will be a major fitness-enhancing characteristic for almost any animal. Cells that can detect light—photoreceptor cells—are a very ancient invention, which predates the Cambrian explosion. A photoreceptor cell on its own can only distinguish light from dark, and even the most primitive visual system needs more than just that. Vision requires the ability to detect the direction that the light comes from, and that in turn requires that the photoreceptor cell be masked on one side, so that it can be used to scan for the light source. In some simple flat worms the eyespots consist of just this sort of two-cell arrangement—a photoreceptor cell plus its masking cell. But if one views visual systems across phylogeny, then an astonishing variety of different forms is found.

In the fruit fly the "compound eye" is composed of hundreds of tiny ommatidia, each of which is equipped with photoreceptors that will provide a colour image of a tiny segment of the visual field. In some molluscs the light enters the eye, bounces off a mirror at the back and then forms the image on a retina that points inwards. We, on the other hand, have a complex camera-type eye, as do cephalopods like the octopus. However, though the octopus eye and the mammalian eye look very similar, they are held to represent a classic case of convergent evolution, because they are built in entirely different ways. The octopus eye develops from an invagination of the embryonic surface layer, while in mammals the eye develops from an evagination of the developing brain. These, and other forms of eye, come and go at an astonishing rate across phylogeny.

In the mid-twentieth century Ernst Mayr reviewed the eye forms known in biology, so as to determine how many times vision had been "invented". By seemingly irrefutable morphological and developmental criteria it was clear that eyes in mammals, eyes in molluscs and eyes in insects couldn't possibly be homologous structures. By these criteria eyes must have been independent analogous inventions that occurred independently between 50 and 70 times. This conclusion, however, turned out to be wrong. And it is wrong because the essence of homology is descent. It is a matter of shared genes and shared genetic circuits, and, as Walter Gehring showed, all of the visual systems known are indeed homologous—if one looks in the right place [3]. To see how this can be, we must first make a short diversion to transcription factors.

E.2 Transcription Factors as Markers of Homology

Transcription factors are one of the prime elements that control the expression of genes. They are proteins, which in the large majority of cases bind specifically to short sequences in the upstream regions of genes and, by doing so, enable the RNA-polymerase to copy the information in the DNA into RNA. Were each of our roughly 20,000 genes to be controlled by a unique transcription factor, then we would need 20,000 transcription factors to do the job, and another 20,000 genes to code for them—and so *ad infinitum*. Things cannot possibly work like this. In fact only around 2000 of our genes code for transcription factors and these therefore have to work in groups. Since groups of transcription factors are the basic unit of gene control, and since one transcription factor can be a member of many different groups, 2000 transcription factors can be used to form a huge number of different controlling groups. Supposing that these factors worked in groups of 4, then 2000 transcription factors could in principle form 8×10^{12} different groups, and this would be more than enough to regulate all our genes in different tissues and in different activation states of the cells. Since any given transcription factor may be part of many different groups, it follows that any mutation which changes its ability to recognise its binding site on DNA will interfere with many different processes and hence will almost certainly be lethal. The result is that natural selection ensures that transcription factors cannot readily change their recognition sequences. It does happen, on rare occasions, but these are very much the exception rather than the rule. A transcription factor in a worm will tend to recognise the same DNA sequence as the homologous transcription factor from a human being.

Which transcription factor is chosen to regulate a gene is a matter of chance for there is no reason to suppose that transcription factor "a" would regulate a gene better than transcription factor "b". However, once transcription factor "a" has been selected as the regulator, it is difficult to switch to regulation by transcription factor b, for all the downstream targets of "a" would have to be individually altered to make them responsive to "b". This is what makes transcription factors such powerful detectors of homology. A trait that is homologous in two species will in all probability be regulated by the same set of transcription factors. A trait that is similar but analogous will almost certainly be regulated by different transcription factors.

E.3 Transcription Factor Cascades

Transcription factors not only work in groups to control individual genes, but also frequently work in cascades, in which a "master" transcription controls the expression of downstream transcription factors, and together they will control the expression of all the genes needed to build a tissue or organ. Loss of the master transcription factor will result in complete loss of the organ or tissue, while loss of any of the downstream factors will often result in only a partial loss of the character. A good example of this sort of thing is seen in the formation of the compound eye in the fruit fly *Drosophila melanogaster*. The transcription factor Pax6 must be expressed in cells of the developing eye, where it switches on downstream transcription factors that regulate all of the estimated 1000 genes needed to form the compound eye of the fly. Pax6 is the "master regulator" for eye formation in *Drosophila*. However, since transcription factors are typically involved in several different genetic circuits it need be no surprise that Pax6 is also required in another genetic circuit which is needed for the proper formation of the head, and in a third circuit which is responsible for ensuring that certain neurones in the brain produce insulin. In mice the development of the eye is also dependent on Pax6. Loss of one allele of Pax6 results in a "small eye" phenotype. Loss of both alleles results in disturbance to the formation of the head and to the development of the insulin producing cells in the pancreas.

The crucial experiment in this story involved the induced expression of the mouse Pax6 gene in cells of the fly that normally do not express Pax6 and instead develop into a leg. The result was a fly with an extra eye on its leg [4]. There are three crucial points here. The first of these is that the Pax6 gene of the mouse was able to organise the formation of an eye in the fly. The second point is that in order to do this, the mouse Pax6 transcription factor had to be able to recognise the appropriate DNA sequences upstream of the genes normally targeted by *Drosophila* Pax6, and it was able to do this despite the fact that the mouse and the fly have gone their separate evolutionary ways for some 500 million years. The third point is that the mouse Pax6 gene in the mouse directs the formation of a camera-type eye, whereas the same mouse Pax6 gene introduced into the fly directs the formation of a fly-type compound eye. Nor is this some exceptional fluke, for Pax6 is now known to be the master regulator of eye formation in all animals with three germ layers—the bilaterians—and an evolutionary precursor of Pax6 directs development of eyes in cnideria.

Gehring and Ikeo explained this in terms of what they called "intercalary" evolution (Fig. 4.10). The idea is that visual systems start with the transcription factor Pax6 being "chosen" to specify photoreceptor cells. Note that there was nothing special about Pax6 that made it peculiarly suitable for this function. The choice was purely random; any old transcription factor would have done the job. However, once chosen, there is no going back, because the targets of the Pax6 transcription factor network needed to form the visual system would have been equipped with the Pax6-binding sites required to ensure their proper expression. The first "eye" would have been a very simple structure, consisting of Pax6 on the one hand and the photoreceptor cells on the other. Since transcription factors recognise rather small DNA sequences of around a dozen base pairs, random mutation will frequently place these binding sites conveniently upstream of genes. Each time that a gene gained a DNA sequence making it responsive to the Pax6 transcription factor network then that gene would be recruited into the eye-building project. If it helped to improve the visual capacity of the eye, it would have been selected and retained. If it did not, then it would not have been selected, and so would have been lost again. In this way different species ended up with quite different structures for their eyes (Fig. 4.10). Structurally or developmentally they may seem very different, but at a deeper level they are all homologous since they are related by descent. Pax6 is the master regulator of them all.

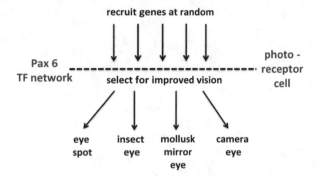

Fig. E.1 Intercalary evolution of visual systems. Starting from a transcription factor network centred on PAX6, which specifies photoreceptor cells, genes may be recruited at random and selected on the basis of their ability to improve the visual system. In different species, different combinations are selected and these different combinations give rise to different eye forms [3]

References

1. Bork P, Holm L, Sander C (1994) The immunoglobulin fold. Structural classification, sequence patterns and common core. J Mol Biol 242(4):309–320
2. Bateman A, Eddy SR, Chothia C (1996) Members of the immunoglobulin superfamily in bacteria. Protein Sci 5(9):1939–1941
3. Gehring WJ, Ikeo K (1999) Pax 6: mastering eye morphogenesis and eye evolution. Trends Genet 15(9):371–377
4. Halder G, Callaerts P, Gehring WJ (1995) Induction of ectopic eyes by targeted expression of the eyeless gene in Drosophila. Science 267(5205):1788–1792

Index

A

Activation-induced deaminase (AID), 48, 78, 79, 98, 103, 105, 112–114, 135
Advantageous mutations, 7, 8, 11
Agnathan, 65, 76–78, 81, 83, 106–115, 131
AIRE, 95, 99
Amoebae, 14, 19, 20, 27, 34, 121
Amphibians, 36, 42, 103, 134
Amphimedon queenslandica, 61, 134
Analogy, 106–108, 138–139
Antibody, 29, 30, 37, 62–66, 80–82, 103, 104, 111, 129
Antigen-binding sites, 83–85
APOBEC3G, 48, 49, 74, 127
Apoptosis, 23, 24, 27, 66, 94, 95, 99, 102
Arms race, 13, 20, 53, 119–131
Autoimmune, 75, 80, 87, 95, 97, 99, 101, 102
Autophagy, 22, 52, 96

B

Bacillus anthracis, 122
B-cell receptors (BCR), 81, 84, 85, 87, 88, 92, 99, 103, 105, 107, 135
Biomphalaria glabrata, 75
Bricolage, 26

C

Caenorhabditis elegans, 4, 26, 27, 36, 73, 74, 120, 134
C3 convertase, 62, 65
CD4, 81, 93–96, 99–102, 105, 114, 125, 127
CD8, 81, 93–96, 100–102, 105, 114, 119, 126, 132

Chemokines, 56, 59, 60
Class-I, 53, 90–93, 100–102, 126
Class-II, 90–93, 96, 101, 105
Class switch, 98, 103, 113
Cnidarians, 61, 65
Coelacanth, 3
Commensals, 15, 27, 44, 54–55
Complement, 46, 47, 56, 59, 61–66, 90, 103, 119, 129
Complementarity Determining Regions (CDR), 82, 84–86, 88, 97
CRISPR, 72, 73
Cyclic GMP-AMP (cGAMP), 49, 50, 137
Cyclic GMP-AMP synthase (cGAS), 49–50, 132, 136–137
Cytidine deaminases, 48, 74, 78, 98, 112–115, 127, 135

D

Danger, 22, 24, 52–53, 55, 101, 136
Danger-associated molecular pattern (DAMP), 136
Daphnia magna, 120
Darwin, C., 2–4, 6, 121
Deinococcus radiodurans, 6
Delbrück, M., 9
Deleterious mutations, 4, 7, 10–12
Dendritic cells, 27, 92, 96, 99, 101, 122
Dictyostelium discoideum, 19, 21
Drosophila, 10, 26, 38–39, 49, 50, 62, 73

E

Ehrlich, P., 28–30, 108

© Springer Nature Switzerland AG 2019
R. Jack, L. Du Pasquier, *Evolutionary Concepts in Immunology*,
https://doi.org/10.1007/978-3-030-18667-8

Endothelium, 59, 60, 66
Exon shuffling, 35, 49, 131

F
Fitness, 3–4, 7, 9, 13–15, 17, 23, 25, 35, 45, 71,
    92, 94, 110, 114, 119, 120, 122, 132,
    138

G
Gene conversion, 78, 79, 83, 98, 103, 105, 107,
    112, 113, 129
Generation gap, 13–14, 67, 110
Genetic drift, 1, 6, 9–11, 13
Germinal centres, 97, 98, 101
Germline, 6–7, 10, 13–15, 18, 28–30, 35, 43,
    53, 67, 71, 72, 74, 75, 79, 84, 113, 115
Gnathostome, 77, 79–82, 84–86, 90, 98, 103,
    104, 106–115, 131, 137

H
Haematopoiesis, 110–111
Helicobacter pylori, 43, 124
Human cytomegalovirus (HCMV), 125, 126,
    128
Human immunodeficiency virus (HIV), 125–
    128

I
Immunoglobulin Superfamily (IgSF), 34, 54,
    76, 77, 81–83, 85, 86, 91, 97, 107, 114,
    135–137
Intercalary evolution, 109, 138–141
Invertebrate, 28, 49, 110, 114, 127, 131, 137

J
Jacob, F., 12, 106

L
Lamprey, 38, 65, 78, 79, 83, 105, 111, 112, 114,
    115
Lectin, 34, 41, 46, 52, 54, 62–64, 136
Legionella pneumophila, 20
Leucine rich repeats (LRR), 34, 37–39, 76–80,
    114, 135, 136
Listeria, 21, 52, 124
Luria, S.E., 9

Lymphocyte, 11, 17, 29, 30, 35, 76–78, 81–83,
    87, 99–102, 105, 108, 110–112, 115

M
Macrostomum lignano, 5, 134
Major histocompatibility complex (MHC), 53,
    89–97, 99–102, 105, 122, 126, 131, 135
Malthus, T., 3
Mannose-binding lectin (MBL), 41, 46, 47,
    61–63, 135
Mavirus, 14, 121
Mayr, E., 1, 3, 139
Memory, 15, 35, 43, 103–104, 127
Microbe associated molecular pattern (MAMP),
    135, 136
Mimivirus, 14, 121
Missing self, 53, 126
Monosiga brevicollis, 61, 134
Moonlighting, 38
Muller's ratchet, 11, 12
Mycobacterium tuberculosis, 120, 125, 127

N
Natural killer cell, 53–54
Natural selection, 10, 67, 75, 94
Necrosis, 24, 27, 66, 90
Negative selection, 94–95
Neutral mutations, 7
Neutrophils, 59, 60
Non-self, 18, 22–24, 46, 48, 51, 61, 62, 64, 65,
    72, 95, 99

P
Pasteuria ramosa, 120
Pentraxin, 34, 47
Phagocytosis, 18–24, 26, 28, 29, 52, 55, 61–62,
    66, 96, 124
Plasmodium falciparum, 8
Positive selection, 43, 93–95
Prokaryotes, 10, 40
Pyroptosis, 66

R
RAG recombinase, 83, 98, 105, 115
Recombination, 1, 11, 12, 18, 36, 37, 41, 78, 83,
    84, 86, 88, 94, 97, 98, 103, 105, 107,
    112, 113, 115, 131
Resistance, 8, 9, 13, 15, 119, 120, 122, 129

RIG-I, 50–51, 135
RNA interference, 73–75

S
Salmonella, 21, 124
Serratia, 120
Shark, 105, 113
Signal transduction, 26, 33, 38, 41, 54–58, 119,
    124
Silent mutations, 7, 8
Soma, 6–7, 14, 18, 72, 74
Somatic mutations, 7, 75
Spätzle, 38
Species, 2–5, 13, 15, 18, 20, 21, 25, 27, 42, 44,
    65, 71, 73, 76, 89, 92, 96, 104–106, 109,
    113, 121, 125, 127, 140, 141
Staphylococcus aureus, 120
Stimulator of Interferon Genes (STING), 49–
    50, 132, 135–137
Strongylocentrotus purpuratus, 36, 134
Symbionts, 15, 22, 27, 54

T
T-cell receptor (TCR), 81, 84, 85, 88, 90, 93,
    96, 100–102, 135, 136
Thymoid, 112

Thymus, 24, 88, 94–96, 108, 112
Tinkering, 26, 49, 76
Tolerance, 51, 75, 80, 87, 94, 97, 99–102
Toll-1, 38–39
Toll-like receptor-4 (TLR-4), 39, 45, 123
Toll-like receptor-15 (TLR-15), 47
Transcription factor, 38, 39, 42, 102, 109–112,
    139–141
Transib, 81–84, 114
T-regulator, 102, 105
Trypanosoma, 125, 128

V
Variable lymphocyte receptor A (VLRA), 38,
    77, 80, 81, 111, 112, 114
Variable surface glycoprotein (VSG), 128, 129
Variation, 1–3, 6, 12, 105, 128
Virulence, 7, 13, 15, 21, 22, 27, 35, 43, 47, 55,
    104, 119–124, 129, 131

W
Whole genome duplication (WGD), 36

Y
Yersinia, 119, 123

Printed in the United States
By Bookmasters